The Art of Architectural Integration of Chinese and Western

Li Haiqing, Wang Xiaoqian

China Architecture & Building Press

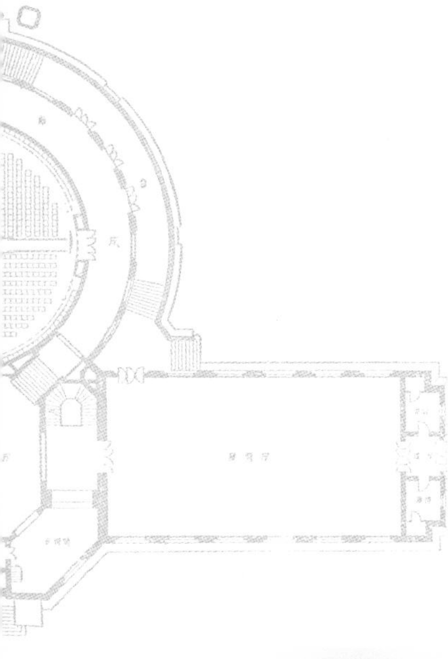

Table of Contents

Chapter 3 Modern Architecture + Chinese Elements

Nationalistic Expression of Zeitgeist

Chapter 4 Folk Wisdom

Folk Culture's Take on Architectural Integration of Chinese and Western

Chapter 5 Significance

Historical Significance of Architectural Integration of Chinese and Western

Preface

Applying dichotomization to "China" and "the West" is amiss, as realized by the perceptive literati. Dissertating about "the architectural integration of Chinese and Western" might be an inopportune undertaking under this context. However, as brawling between Chinese medicine and western medicine, Chinese cuisine and western dish, Chinese painting and western painting happens almost every day, we cannot help but have to acknowledge perceptively the cultural difference conspicuously exists and impacts our daily lives down to minute details.

In a global context, however, the emergence of modern nation states since the 16th century had ended the era of long-term isolated survival and development of nations. Inspired by thriving seaborne trades, the eastward navigation of Portuguese influenced Chinese architectural culture, resulting in the frequently mentioned dynastic division and correlation in the modern world history, modern history of China and Chinese modern architectural history. Regardless of the division of time lines, the earliest cultural collision was affirmatively from European (the Occidental) although relayed from the colonial causes in South and Southeast Asia: dominated by the modulated occidental architectural culture, viz. the "Veranda Style". A few centuries afterwards, the eventful modernization movement in the world history was sprouted by the instigators in their own region, namely, the west Europe and Northwest Europe. Therefore, this modernization and the westernization were synonyms. The division of "China" and the "West", or the "East" and the "West" appeared in the international political culture context. This explains the later movements in the East of the "Westernization with Chinese Characteristics" and "Wakon Yosai". This is attested by the statement of Samuel P. Huntington: "Chinese civilization is the longest civilization of the world. The Chinese are conscious about the peculiarity and achievement of their civilization. The Chinese literati tackle naturally world issues from a civilization perspective. They realize the world has diverse civilizations." Nevertheless, learning and collaboration coexist with competition. The past six hundred years were a period of close interaction and learning of world's civilizations. The frequency and depth have been enhanced by time. The emergence of modern nation states and acceptance of the "imaginative community" have facilitated the increasing intensity of contention and collaboration influenced by the political Machiavellianism, which led to incontrollable chaos twice—two unprecedented world wars in the 20th century. At least up to now, the final victory belonging to just or political might remains inconclusive. The process of the clash and exchange of architectural cultures was ignited under the above background, which cannot be isolated.

It has been nearly 20 years since the approval of the authors started their research of the modern and contemporary history and theories of Chinese architecture. During this period, the trend of the study of architectural theories has shifted from the interpretation of architectural forms, spatial construction to construction analysis. However, no matter what perspective taken toward the exploitation of materiality and technicality of architecture, culture-centric dynamics cannot be ignored, since technology is a manifestation of culture. The conceptual system formulated under cultural environment is the key to understand the multitude reasons of architectural activities. Consequently, it inspired the publication of the *Study on the Typological Modernization of Chinese Architecture* of Southeast University in 2004 and the study on the career system of architects in modern China. In the last decade, another inspiration stemmed on architectural implementation triggered a new round of contemplation under the transformation context of "dual industrialization" and the realistic circumstance of "semi-industrialization", which have brought up the discussion to constructive culture.

Observing such elusive pivoting, the authors could not resist to the glamour of *Superimposition and Reconciliation——Art of Architectural Integration of Chinese and Western*. To traditional Chinese intelligentsia, Confucianism's serving the Great Way, Taoism's nourishing life, and Buddhism's nourishing mind can coexist without revolt on the same living being. This is an ingenious integration of ideology, disposition and cultivation. It is so-called the "Conjunction of Three Creeds into One". Whereas, serving the Great Way is too challenging and ideal that it has become "sophisticated egoism" under the inducement of instrumental rationality. This is exactly the issue we must face, revisit, contemplate and discuss: What is our soul? How to cognize other's value? How to forejudge our path?

The authors are grateful for the planning of China Architecture & Building Press and the recommendation of Professor Chen Wei. Thanks to them, we have the opportunity to do some summary and analysis of the previous studies. In addition to the overall planning of the entire book, Li Haiqing is also responsible for writing Chapters 1, 2 and 5 and Wang Xiaoqian completes Chapters 3 and 4. Advice or comments from experts of various fields are welcomed.

Li Haiqing, Wang Xiaoqian
Nanjing July 2014

Chapter 1 Overview

Contemporary Architectural Integration of Chinese and Western in Chinese Architectural Art

In the early spring of the Year of the Dragon, east wind thawing, while had the central government been tightening policy of curbing housing price over a year and had Chinese housing market been sluggish and indolent, a significant event snapped the dreariness. On February 29, 2012, Architecture BBS (ABBS) cited a report of British Broadcasting Corporation (BBC), Chinese Architect Wang Shu was rewarded the 2012 Pritzker Architecture Prize and became the first architect of a Chinese citizen who has ever received such a top award in the architectural field. Ieoh Ming Pei, also a Pritzker Laureate in 1983 for the similar achievement, is a Chinese American, which signifies Wang's laurel. Thomas J. Pritzker, chairman of the Pritzker Architecture Prize and the Hyatt Foundation elaborated, "This is an epoch-making leap. The final decision to laureate a Chinese architect by the jury solutes the Chinese architecture advancement with international recognition." This is a remarkable affirmation of the advancement of Chinese architecture by the rest of the world. Confirming the possibility of exploring "China Problems" from the architecture-independence perspective, Wang's works have been described as "the ingenuity of Chinese empirical construction system of regionalism", where China, regionalism, folk and construction are keywords, delivering profuse and comprehensive information yet narrating with personalization and plebeian attributes. That are even given with the same praise to Russell—the greatest invention—would not be insipid or loutish.

To the rest of the world, "China" had become an international issue after the successful global navigation in the 16th century, especially conspicuous since the Industrial Revolution of modern time. From academic view, any historian will agree what Chinese inevitably have to confront has started since the 1840s—"unprecedented transformation", and continues. The impact beyond elucidation is deepening and spreading. The shift from a traditional "agriculture society' to modern "commercial and industrial society" should be the most phenomenal change in the past 170 years of China's history. Under the prospect of social transformation, the aspects of economy, politics, military, culture, education and medical care have reacted accordingly. As an important cultural element, the art of architecture resonates in synchronization as well.

The reorientation of the architectural art has been focused on two aspects. One is the "migration" of western civilization, such as the ferroconcrete construction technique, handicraft and artistic expression of "modern architecture", which were not elements in Chinese traditional architectural culture, and any then extant knowledge and experience when Shanghai Hongqiao Nursing Home (Fig. 1-1) was built in the 1930s. Another one, rather being "grafting" not "migrating", connecting more or less with the traditional Chinese architectural culture, profoundly or superficially, in simple or complicated artistic styles, as a result of coactions of the two cultural memes, features neither Chinese nor western, or harmonized Chinese and western attributes. As for the "grafting", the categorization of Chinese-Western integration, Chinese folk-Western integration, and Chinese-Western confluence are for the purpose of appraisal. Among them, the most testimonial category is the "Chinese-Western" architectural integration in which Guangzhou Sun Yat-sen Memorial Hall (Fig. 1-2) and Nanjing National Government's Foreign Ministry (Fig. 1-3) built in the 1930s are the most prominent exemplars of this style.

Fig. 1-1 Shanghai Hongqiao Nursing Home

Fig. 1-2 Guangzhou Sun Yat-sen Memorial Hall

Fig. 1-3 Nanjing National Government's Foreign Ministry

1.1 Exemplars of the Art of "Architectural Integration of Chinese and Western"

1.1.1 Concept of "Architectural Integration of Chinese and Western"

Cihai (a term dictionary) explains, "Integration of Chinese and Western" as

He Bi: A *Bi* is a flat jade disc with a circular hole in the center. A semi-circular *Bi* is called *Ban Bi*. "He Bi" is a circular *Bi* formed by two pieces of *Ban Bi*.

An analogy can be given as the combination of two meritorious scenic spots or buildings of Chinese and foreign into one.

Source: Cited from the *Marvels of Officialdom*, Li Baojia (the Qing Dynasty), "Today, we'll try to wed Chinese and Western ...the main seat is next to here; Mr. Sa sits on the right and his company, Mr. Liu takes the left seat."

Synonym: Chinese folk-Western integration, Sino-western.

Conspicuously, the "integration of Chinese and Western" is a confirming annotation of this unique phenomenon of cultural collision of Chinese and Western in that historic time. Its premise is:

a. Requirement of "Ingenuity": Regardless of the original sources, the ingredients integrated must be the essences, or the so-called "Bi";

b. A requirement of "Style": While from heterogeneous elements, similar to "He Bi", the integration must be flawless.

Therefore, from the aesthetic significance, "He" of the integration of Chinese and Western means harmonization or coherence but an "admixture".

The English translation for "Chinese-Western" architecture is "Architectural Integration of Chinese and Western".

1.1.2 Principle and Formation Mechanism of Architectural Integration of Chinese and Western

Factors influencing the artistic finesse and relish of the architecture are complicated. By exploring into the circumstances of the construction, the function and space, techniques/materials, site/environment and shape/style of the building stay as key parameters that manifest the infusion of interactive collision of different architectural cultures. The manifestation includes the Chinese-Western coherence in the same element or Chinese-Western grafting of different elements. For example, the "Contemporarily Styled Chinese Classical Architecture" is such typical case, which is a unique form retaining the architectural format/style of the

Fig. 1-4 Sun Yat-sen's Mausoleum, Nanjing

ancient Chinese official building and grafting it with modern western construction techniques/materials in almost even weights of "Chinese" and "Western" attributes. It is different from "Chinese Modern Architecture" that adopts completely the western function/space and techniques/materials, only with a slight touch in detailed decorations imitating the traditional Chinese wood architecture on modern building blocks in form and style. This approach belongs to the Chinese-Western coherence in the same element, with emphasis more on "Western" attributes of the two.

1.1.3 Types of "Architectural Integration of Chinese and Western"

From perspectives of architecture principles and formation mechanism, reviewed from the construction environment of historical relics of the artistic style of edifices, the architectural integration of Chinese and Western can be categorized into three types:

1. Contemporarily Styled Chinese Classical Architecture

The new genre, the antique palatial style and a hybrid style of eclecticism[1], grafts the modern construction techniques/materials onto the format/style of ancient Chinese official buildings. Its examples include administration buildings, cultural education centers, and assembly and memorial and religious edifices constructed by the government and professional groups. Usually, they are large buildings with impressive investment, e.g., Sun Yat-sen's Mausoleum, Nanjing (Fig. 1-4) and Shanghai Special Municipal Government building (Fig. 1-5).

2. Chinese Modern Architecture

This type of architecture, fashioned modernism with heavy decorations[2], is infused completely with attributes of function/space and techniques/materials of the western architecture and detailed decorations imitating the traditional Chinese wood architecture to fashion the modern building blocks in form and style, i.e., embellished with "Chinese Elements". Those buildings, designed by professional architects, include administration buildings, cultural education centers, residential mansions and large functional public buildings, as well as assembly, performing centers,

1 Hou Youbin (2001), Form and Tide of Thought in Architecture, In Pan Guxi (Ed.), *Chinese Architecture History* (4th ed., pp 382-385), Beijing: China Architecture & Building Press

2 ditto

Fig. 1-5 Shanghai Special Municipal Government Building

Fig. 1-8 Folk houses in Wuhan "Lifen"

Fig. 1-6 Main entrance to the National Central Hospital, Nanjing

Fig. 1-7 Folk houses in Shanghai "Shikumen" (or Stone Warehouse Gate)

and medical buildings, such as Nanjing National Government's Foreign Ministry (*also see* Fig. 1-3) and the National Central Hospital (Fig. 1-6).

3. Buildings Invested by Private and Non-Professional Groups

Most of these buildings, with non-governmental or much less ministerial involvement, built by private owners, reveal more convoluted conjunctions in perspectives of function/space, construction techniques/materials, site/environment and genre/style. They are not only in forms of urban alley dwellings, different from traditional folk houses, such as Shanghai "Shikumen" (or Stone Warehouse Gate) (Fig. 1-7), Wuhan "Lifen" (Fig. 1-8) and Tianjin "Garden-Style Lilong" (Fig. 1-9), private manors and fort villages with copious features and combined merits of genres, such as the Huzhou Nanxun Jiayetang Library (Fig. 1-10), Suzhou Dongshan Building of Carvings (Fig. 1-11) and Kaiping Watchtowers(Fig. 1-12), but also more small and medium-sized commercial buildings with "western-style façade" spread across urban and rural areas, such as shops of the Ruifuxiang Silk Store in Beijing (Fig. 1-13).

The secular inclination and dignitary favoritism are two weighed attributes in Chinese architecture culture that orient the aesthetic disposition and interest of the public toward the architectural form/style rather than the function/space and techniques/materials. To wit, occupied by the "appearance" (merely visual symbolicalness of semiotics), people are indifferent to the "nature" (the actual function of the building

Fig. 1-9 Folk houses in Tianjin "Garden-style Lilong" (from Zhang Wei, Tianjin University)

Fig. 1-10 Huzhou Nanxun Jiayetang Library

Fig. 1-12 Kaiping watchtowers, Guangdong

and the key social, economic and environment attributes of the architectural performance) and the "background" (architectural activities as the kinetic model of social, political and economic resource allocation and integration) of the architecture. This induces this building genre arising from the private sector as the typical and popular visual art of the "architectural integration of Chinese and Western". Professional architects with profound understanding of architectural theory are easily obsessed by the nature and background not the appearance in the opposite to the non-professionals who merge generously all possible artistic means in the creation of visually appealing buildings.

Fig. 1-11 Suzhou Dongshan Building of Carvings (from Shi Xing, Southeast University)

Fig. 1-13 Ruifuxiang Silk Store in Beijing (provided by CFP)

1.2 Overview of "Architectural Integration of Chinese and Western"

1.2.1 Background

To discuss the background of the "architectural integration of Chinese and Western", it is inevitable to face the subject of the clash of civilizations.

Since the 1840s, the ancient Chinese civilization has been experiencing the unprecedented fierce foreign cultural impact. Chinese architecture, as one of the cultural elements, has inescapably been forced to make tremendous changes. Facing centuries of, yetin new forwarding redirection—the global modernism, Chinese architecture has no exception but to move along with that of China and Chinese people. American scholar C. E. Black had pointed out: There are three great revolutionary transformations in human history. The first time happened one million years ago when the primordial creatures evolved for 100 thousand billion years into human race. The second time was when humans created civilizations. The third time was the recent centuries when the entire human race transformed from the agrarian/pastoral civilization into the industrial civilization. Evidently, this transformation has different characteristics than those of the previous two: First, from the globe perspective, the previous two transformations were regionally and ethnically isolated and occurred independently. The third one was entirely different. The inevitable dispersion and expansion characteristics of non-stop modernization movements of the industrial revolution, science and technology advancement, information explosion and information revolution, contributed and facilitated by convenient transportation, telecommunications, were also accompanied by confrontation, conflict and blood shed amid aggressiveness and coerciveness, while, in the opposite, opened ways for reciprocity and exchange among civilizations.[1]

Regionally, the modernization transformation process ostensibly imbuing with Western European taste that perforating continents of America, Australia, Asia and Africa along with the colonization, referred to as Occidentalization or Europeanization, conveyed fecund undertone. The modernization was not a universal or simply an acknowledging and attaching gesture toward the Occident. As a reaction, deviations in value orientations and approaches taken by countries and ethnics according to individual historical cultural trait and favoritism could not be avoided. To the ancient Chinese culture that has salience in the world civilization history, the circumstance was more convoluted not merely a simple issue of "Occidentalization" – the macroscopic perspective of historical elicitation of the "harmonization of Chinese and Western".

Because of the dispersibility and expandability nature of the third revolutionary transformation of the human kind, the starting point and reaction of modernization processes varied by countries. Based on that, they can be categorized as the "Early Endogenous Modernization" and the "Late Exogenous Modernization". The former includes Britain, France and the United States, whose modernization processes began as early as the 16th—17th centuries and started from their inner societies. The latter includes German, Russia, Japan and many current developing nations, whose modernization processes started as late as the 19th century, sparked by external confrontations of survival and encouraged by examples of the "Early Endogenous Modernization". Evidently, China's modernization fell into the latter case.

To the reaction subject and the cross-pollination mechanism, the "Late Exogenous Modernization" is a more complicated process, which makes it more fascinating. To the traditional Chinese civilization and its conservatives, after experiencing the initial benightedness and feeble resistance, the overall mentality toward the impact of the western civilization was rather paradoxical in one way admiring its advance and wanting to import and exploit it, and in another, worrying its prevalence and imposing

1 Xu Jilin & Chen Dakai (1995.1), *Modernization History of China* (Vol. 1, pp 1800-1949). Shanghai: Shanghai Joint Publishing Company

Fig. 1-14 One Western mansion of the Gardens of Perfect Brightness

restriction and control. By compromising, doping Chinese elements into the western-style architecture or adding western architectural elements into the traditional Chinese architecture became the two basic approaches, creating the genre of the "architectural integration of Chinese and Western". More interestingly, from the evidence of modern Chinese history, the initiator of the "architectural integration of Chinese and Western" was not the defending Chinese but the intrusive westerners—Christian missionaries and the church charging in China.

1.2.2 Progress Overview

The progress of the "architectural integration of Chinese and Western" can be divided into two periods, the Late Qing Dynasty and the combined period of the Republic of China and New China. The latter can be further divided into three stages, the Prelude (1800s—1920s), the Prime (1920s—1930s) and the Descent (1940s) stages.

The activists in the Prelude stage were the western churches and western architects. Earlier than the time of the Opium War, the Qing Imperial Court hired missionaries Giuseppe Castiglione, R.Michel Benoist and Jean Denis Attiret to preside over the construction of Western mansions of the Gardens of Perfect Brightness (Fig. 1-14), which is the most significant case of the "architectural integration of Chinese and Western" in China's pre-modernization time. However, its imperial garden nature limited its influence on the society. After the Opium War, Christianity (including Catholicism, Protestantism and Eastern Orthodox) regained their legal status in China. However, different from the past, this time, the migration of "foreign religions" caused malicious woes often between the church and Chinese populace. The conflicts and paradoxes triggered the "Populace-Religion Conflict", namely, the "Ecclesiae Incident". The Incident is not a modern phenomenon and it can be traced back to the similar incident brought by the missionaries of the Society of Jesus during the late Ming Dynasty. Both were rooted to a cultural background with immense populace support, during which adherents of Confucianism, Buddhism and Taoism were the main resistances against the Catholic. What more crucial was the strong political insinuation of the invasive coherence with the western colonialists. The frequent occurrence of "Ecclesiae Incident" foiled the religious mission that persuaded the churches to adjust their course as inspired by the experience of Matteo Ricci, et al., by harmonizing with Chinese culture—following Chinese dress code, preaching in Chinese and even more significant measures: "Sinicization" in forms of buildings of churches, schools and hospitals.

Fig. 1-15 The Holy Saviour's Cathedral, Nangouyan, Beijing (provided by CFP)

Fig. 1-16 Interior of the Holy Saviour's Cathedral, Nangouyan, Beijing (provided by CFP)

Thus, "Chinese-styled" buildings emerged. In the early 20th century, the compromise finally formed the "Sinicization" of the Catholic and the "Localization Movement" of the Protestant. It created the integration of the western church structure and the traditional Chinese wood roof, such as Holy Saviour's Cathedral (the former Anglican cathedral in Xicheng District of Beijing, China) (Fig. 1-15, Fig. 1-16) and the campus building of Ginling Women's College of Arts and Sciences (a Christian university founded in 1913 in Nanjing, China) (Fig. 1-17).

Meanwhile, the populace gradually redirected their gainsaying to crave and admire the western civilization and under such influence, simple attempts applying the "western-style façade" to merchant buildings by marginal groups had obtained success.

Fig. 1-17 Ginling Women's College of Arts and Sciences (a Christian university founded in 1913 in Nanjing, China)

The approaches were smartly done by building a two- or three-story brick wall. The shop front was rendered with carved decors, finishing the top extrados of door or window openings with straight arch or chamber arch, adding carved embellishments on the capitals of pilasters to deliver a variant of similar architecture of European Renaissance, and topping parapets paneled with motifs in arch or peach-shaped design. In general, by simply using a full-bodied façade wall, a building with a taste of the western style was achieved efficiently. The "Eastern-Western" art can be characterized as: a) retaining the horizontal inscribed board, shop sign and couplets made of enforced paper or hemp-lime mixed mortar on the shop front façade; b) adding carved decors of the "western-style façade" with traditional Chinese subjects, such as "Dragon and Phoenix Proffer Bliss", "Kylin sending Babies to the Mortals" or "Three Friends of Winter". They were framed with iron wires or bamboos and cast with hemp-lime mixed mortar before being variegated. For example, the Ruifuxiang Silk Store located at Beijing Qianmen Dashila (*also see* Fig. 1-13) was built in 1893 and rebuilt in 1900 after destroyed by a fire. It has two-story brick-and-wood structure with the façade in dark green color and white marble carving decors of "Pine and Crane Bless Longevity", "Peonies Present Wealth and Nobility" and "Lotus Praise Integrity and Virtue". In general, it is a typical traditional folk design touched with only some western variant details on the façade. It can be referred to as "an exemplar of the manifestation of foreign architectural culture adopted by Chinese civil society".

The Prime stage started at the end of 1920, the protagonist was a Chinese architect who just created a new chapter in Chinese architectural history. After declarations of the victory of the Northern Expedition and a unified regime of China by Nanjing National Government, the Kuomintang (KMT) government announced the "Political Tutelage" period. To empower its mobilization supremacy of the entire society, it asserted the renaissance of the traditional Chinese culture. Marked by the *New Chinese National Capital Planning* and *the Greater Shanghai Plan*, "China's inherent architectural style" regained its predominance. Its core ideology was to preserve "merits of Chinese art"[1] on the basis of "Western scientific principles", which was elaborated as "to preserve Chinese design is to incorporate only the refined Chinese excellence and to integrate the foreign exquisiteness. So that Chinese design is adopted for the exterior as the main course and the foreign for the interior as the side touch."[2] Architect Lü Yanzhi rising to his fame for designing Sun Yat-sen's Mausoleum and Guangzhou Sun Yat-sen Memorial Hall also said, "Today, China is renewed. The governance is restored anew. All public buildings reflect the spirit of the new construction of the people. They must be built with characteristics of Chinese style with profundity, artistic patterns and scientific structure, heralding the prospect of Chinese architecture."[3]

Impelled by the nationalism ideology, most of the public buildings under the governance of the National Government were designed with the "traditional Chinese curved roof"—"Chinese architectural quintessence" and ferroconcrete or steel structure—"scientific principle". The "architectural integration of Chinese and Western" manifested in this time was the solemn and magnificent appearance of Chinese official buildings with the combination of the advanced architectural technique and construction technology, such as Guangzhou Sun Yat-sen Memorial Hall built in 1931. It has a 4,600 seats auditorium topped with octagonal pyramid roof supported by using 30-meter span *Fink* steel roof truss (Fig. 1-18). Driven by the policy, buildings with "China's inherent architectural style" constructed in this period, just in Nanjing alone, were 30. Other cities, such as Shanghai, Guangzhou, Wuhan and similar places, had also a stunning number of new buildings in this style. Even, Henan Kaifeng, Xinxiang, and so did other secondary cities. Some high educational institutes followed the trend and had their new buildings designed in the same fashion, such as campus buildings in Sun Yat-sen University by Yang Xizong and Lin Keming, campus buildings in Lingnan University, Fudan University and Ginling Women's College of Arts and Sciences by Henry K. Murphy, campus building of Wuhan University by F. H. Kales, and campus building of Xiamen University by Chen Jiageng, etc.

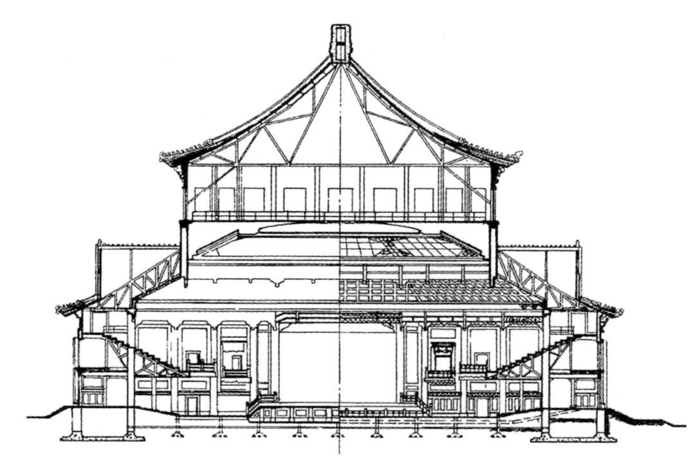
Fig. 1-18 Fink steel roof truss of Guangzhou Sun Yat-sen Memorial Hall

1 Sun Ke (1929), Preface to New Chinese National Capital Planning, *New Chinese National Capital Planning*, Nanjing: New Chinese National Capital Design Technical Specialist Office

2 *New Chinese National Capital Planning* (1929, pp 35), Nanjing: New Chinese National Capital Design Technical Specialist Office

3 Lǚ Yanzhi (1929), *New Chinese National Capital Urban Metropolitan Area Design Draft Drawings* (Set 1, pp 25), New Chinese National Capital Design

Fig. 1-19 The National Great Hall of the People, Nanjing

Fig. 1-20 The National Art Gallery

Meanwhile, affected by the western "Modernism" and the criticism of high construction cost of "China's inherent architectural style", another newfangled genre "Chinese modern architecture" had sprouted. It took on architectural function/space and techniques/materials completely from the contemporary western architecture with simple geometrical shape and flat roof, and was built in ferroconcrete alone or mixed with brick, imitating detailed decors of the traditional wood architecture only in form/style. Its innovation of blending a few traditional Chinese architectural elements in the modern architecture design was an unprecedented approach. Examples for this genre include Nanjing National Government's Foreign Ministry, the National Central Hospital in Nanjing, the National Great Hall of the People in Nanjing (Fig. 1-19), the National Art Gallery in Nanjing (Fig. 1-20) and Headquarters of the Bank of China in Shanghai (Fig. 1-21). The latter, designed by Chinese architect Lu Qianshou, was the first high-rise architectural design of the "architectural integration of Chinese and Western" style.

Another compatible "architectural integration of Chinese and Western" style differed from those of official, large-scale public edifices. The construction activities directed by architectural professionals were alley dwellings—built in large numbers earlier at treaty ports, such as Shanghai, Tianjin and Wuhan, and private manors and fort villages, which were not vast in the number but unique in features and in combined merits of different genres, such as the trendsetting private gardens in exuberant Jiangnan (regions, south of the Yangtze River) and nothing-in-common Kaiping Watchtowers, which had building style of the hometown of overseas Chinese in the Lingnan region. The generality of these is the design and construction of the "architectural integration of Chinese and Western" being self-motivated, planned and funded

Fig. 1-21 The Headquarters of the Bank of China in Shanghai

by the owner or real estate builders, participated by Chinese construction contractor and craftsman in the design and implementation processes. Because of the non-professionalism factor, the style, taste and purpose of the architectural art were diversified and exuberant. For example, Suzhou Dongshan Building of Carvings built in 1925 (*also see* Fig. 1-11), which is rich in delicate carvings, amazing structure and named "The Best Building in Jiangnan". The luxury mansion was built by Jin's brothers who became successful merchants in Shanghai. It was designed by a well-known architect Chen Guifang and constructed by over 250 artisans. It cost 150,000 Chinese silver dollars, and it took three years to finish. Taizhou Gaogang "Building of Carvings" built in the early republican period has three connected arch gates in its traditional Chinese courtyard with "Western Gate" intaglios incised on the fan-shaped pattern of the overhead board intentionally indicating its foreign cultural source. The inner doorframe has traditional Chinese patterns of golden pheasant, magpie and twin deer accompanied by foreign elements of a lion crouching on a terrestrial globe and a flying horse with wings. The "architectural integration of Chinese and Western" expressions permeate conspicuously.

However, the Prime period of the "architectural integration of Chinese and Western" did not last long. The twelve-year chaotic warring state from 1937 to 1949 brought "China's inherent architectural style" to its downfall. In this period, the national economy was under the wartime restraining order. "China's inherent architectural style" suffered fewer interests due to its massive cost and unsuitability to the new economic environment. Only a few were built, such as the library (Fig. 1-22), physicochemical Building and student dormitory in Wangjiang Campus of Sichuan University, Chengdu, Chengdu Liu Xiang Cemetery, and the Administration Office of the Academia Sinica, Nanjing (Fig. 1-23) designed by Yang Tingbao, the Victory Memorial Hall of Anti-Japanese War in Kunming (Fig. 1-24) by Li Hua. In the Descent period, "China's inherent architectural style", to save financial investment, was forced to use domestically available materials, such as the wood roof trusses

Fig. 1-22 Library in Wangjiang Campus of Sichuan University, Chengdu

Fig. 1-23 Administration Office of the Academia Sinica in Nanjing

Fig. 1-24 Victory Memorial Hall of Anti-Japanese War in Kunming

Fig. 1-25 Sun Yat-sen Hall in Wugang, Hunan

Fig. 1-26 The Cultural Palace of Nationalities, Beijing

for campus buildings of Sichuan University. Despite the wartime, the westward relocation of the National Government stimulated the economic development and the city construction of China's mid-west region. The "architectural integration of Chinese and Western" style architecture activities in this region were geared up, such as the Sun Yat-sen Hall in Wugang, Hunan built in 1943 (Fig. 1-25). The hall adopted the single layer hip roof design, using mixed traditional Chinese column-lintel structure and column-crossbeam-lintel structure to support its roof, while mixed with western brick-wood for its body structure, triangular pediment gatehouse in the front facing façade, delivering some western classical architecture taste. In addition the vagrant state of endless warring chaos also inspired architects to realize the value of the traditional house design of common dwellings of Chinese inhabitants. Their design ideology and approach deserve more serious studying and learning. As a result, the Yingqiu Court (or Autumn Muse Court) by Lin Huiyin becomes the typical case.

The short period of the National Government ended in 1949. The establishment of New China and the new prevalence of the political system could not sever the cultural linage. The "architectural integration of Chinese and Western" style buildings, after 1950, embraced another peak time. The "nationalism with socialist characteristics" imported from Russia pushed "China's inherent architectural style" to a new stage. Of course, together with "Chinese modern

Fig. 1-27 Beijing Friendship Hotel

Fig. 1-28 Beijing Railway Station (provided by CFP)

Fig. 1-29 The Administration Building of the Chinese Ministry of Construction, Beijing

Fig. 1-30 Chongqing People's Auditorium

architecture", it was packaged into a "nationalistic style". Under the guidance of "applying historical merits to modern applications, transforming western merits into Chinese instruments, and innovating to renew the traditional value", exemplified by "Beijing Ten Great Buildings", the "architectural integration of Chinese and Western" genre created another batch of masterpieces. Among the designs, many were derived from, characterized by "China's inherent architectural style", and expressed with the "traditional Chinese curved roof" in various formats, degrees and emphases, such as the Cultural Palace of Nationalities in Beijing (Fig. 1-26), Beijing Friendship Hotel (Fig. 1-27), Diaoyutai State Guesthouse in Beijing, Beijing Railway Station (Fig. 1-28), the National Agriculture Exhibition Center and buildings of four ministries and one national assembly hall[1]. There were some cases continued the "Chinese modern architecture"

Fig. 1-31 The Teaching Building of Eastern China Aeronautics Institute in Nanjing

1 "Buildings of Four Ministries and One National Assembly Hall" collectively referred to as the massive administration building complex in then Beijing, were designed by the known architect Zhang Kaiji and built in 1954 by a joint construction task force of the State Planning Commission, Ministry of Geology, Ministry of Heavy Industry, First Ministry of Machine Industry, and Second Ministry of Machine Industry during the early period of PRC. See: Zhong Da (1955), *"Four Ministries and One National Assembly Hall"—Office Buildings from a Cost-Effective Perspective View*, Architectural Journal (Issue 1)

Fig. 1-32 Lu Xun Memorial Hall, Shanghai

Fig. 1-33 The National Library of China, Beijing (provided by CFP)

Fig. 1-34 Beijing West Railway Station (provided by CFP)

exploration efforts, such as the Administration Building of Chinese Ministry of Construction, Beijing (Fig. 1-29). Synchronizing with the movement, major cities joined the fads, for example, Chongqing People's Auditorium (Fig. 1-30) and the Teaching Building of the Eastern China Aeronautics Institute in Nanjing (Fig. 1-31), of which there are some successful works that were designed in the style of traditional common dwellings, such as Lu Xun Memorial Hall, Shanghai (Fig. 1-32). In the following decade, the enthusiasm faded with the economic policy and ten years of civil chaos.

The reform and opening-up in the late 1970s provided architecture another rare opportunity. In addition, the cross-stimulation of "Postmodernism" in vogue internationally offered a normalization stage for an overall exploration, refinement and application of China's architecture cultural heritage with compromise and integration with the foreign advanced architecture technology, model and form. Encouraged by the gradually relaxed political atmosphere and the introduction and study of modern architecture theories and ideas, this normalization was pushed into a new historical period of deepening and expanding internally to the architectural profession, during which, the new National Library of China (Fig. 1-33), Beijing West Railway Station (Fig. 1-34), Qufu Queli Guesthouse, Shandong Province (Fig. 1-35), Shaanxi

Fig. 1-35 Qufu Queli Guesthouse, Shandong Province

Fig. 1-36 Shaanxi History Museum

Fig. 1-37 Yuhuatai Revolutionary Martyrs' Memorial Hall in Nanjing

Fig. 1-38 Wuyi Shanzhuang Hotel in Fujian

Fig. 1-39 The Fragrant Hill Hotel, Beijing

Fig. 1-40 Fengzeyuan Restaurant, Beijing (provided by CFP)

Fig. 1-41 Pan Tianshou Memorial Hall in Hangzhou

History Museum (Fig. 1-36), Yuhuatai Revolutionary Martyrs' Memorial Hall in Nanjing (Fig. 1-37) and Wuyi Shanzhuang Hotel in Fujian (Fig. 1-38), built in succession since 1980 belonged to the best group in this period. While, outstanding design works like the Fragrant Hill Hotel, Beijing (Fig. 1-39), Fengzeyuan Restaurant (Fig. 1-40), Pan Tianshou Memorial Hall in Hangzhou (Fig. 1-41), Zhejiang Art Museum (Fig. 1-42 and 1-43) and Vanke Fifth Park, Shenzhen (Fig. 1-44), are new rises of "Chinese Modern Architecture".

Not just in Chinese Mainland, the "architectural integration of Chinese and Western" found its niche in Taiwan since 1950, where, the Taipei Sun Yat-sen Memorial Hall (Fig. 1-45), the Grand Hotel in Taipei (Fig. 1-46), the Chiang Kai-shek Memorial Hall in Taipei (Fig. 1-47) and Chinese Culture University (Fig. 1-48) are representatives for this period.

The year of 2009 was the 60th national anniversary of the People's Republic of China, elicited by the grand and solemn ceremony settings and ardency of celebration of the entire nation, *Architectural Journal*, the official and authoritative professional architecture periodic, prompted the "Architecture Talents"—photos of winning awards of the National Architectural Design. From the set, if viewing carefully, at least over a dozen of 50 photos on the front cover and back

Fig. 1-42 Zhejiang Art Museum, Hangzhou

Fig. 1-43 Interior of Zhejiang Art Museum, Hangzhou

Fig. 1-45 Taipei Sun Yat-sen Memorial Hall

Fig. 1-44 Vanke Fifth Park, Shenzhen

Fig. 1-46 Grand YuanshanHotel in Taipei (from Guan Hua, Nanjing University)

Fig. 1-47 The Chiang Kai-shek Memorial Hall in Taipei

Fig. 1-48 The Chinese Culture University in Taipei (from Guan Hua, Nanjing University)

cover were buildings of the "architectural integration of Chinese and Western" style, incorporated with the "nationalistic style" and "Chinese elements". The evaluation results published at this time drew some attentions. For the overall presentation, the designs were affected by the "Late Exogenous Modernization" model and "Chinese Cultural Renaissance" movements. They struggled to catch the global trend and preserve the traditional cultural attributes with hesitation amid a centurial teetering Chinese architecture industry, whereas the "architectural integration of Chinese and Western" genre created under this convoluted and ambivalent mentality has formulated the resplendent background for China's contemporary architecture.

Chapter 2
Contemporarily Styled Chinese Classical Architecture

Eclecticism and Virtuosity

2.1 Christian Localization and Church Buildings

2.2 Experiments of Ingenuity: Nanjing Sun Yat-sen's Mausoleum and Guangzhou Sun Yat-sen Memorial Hall

2.3 Nanjing National Government and "China's Inherent Architecture"

2.4 Reassessing College Campus Architecture

2.5 New China: "Traditional Chinese Curved Roof" with "Nationalism Form Integrated with Socialism Content"

2.6 "Cold War" Taiwan: "Chinese Cultural Renaissance" and Architecture

The ideology of creating an architectural genre with "Chinese characteristics" by implementing advanced western scientific architectural methods has been effervescent in China's contemporary architectural activities. It glowed in various formats in the past. Regardless of names imposed on it, from the "Sinicized" church buildings of 1920s to "China's inherent architectural style" in the 1930s, followed by the "nationalistic style" of 1950s and 1960s and the "New China style" of 1980s, its theoretical basis and practicality pursuit stayed unchanged. The art of architecture is a cultural attribute of a nation and its people, and its expression is the testimony of the very quintessence of the attribute. Therefore, Chinese architecture must present its own core nature. In addition, the ideology is coherent with the adoption of the advanced western scientific architectural means in materials and technology perspectives. In fact, the marriage of the two will reinvigorate Chinese architecture – Fu Chaoqing (a professor of Architecture) has given it a subtle name as "Contemporarily Styled Chinese Classical Architecture"[1].

The "Contemporarily Styled Chinese Classical Architecture" has at least two core elements: One is the conformational characteristics of the ancient Chinese wood official buildings, such as the "traditional Chinese curved roof", imitated wood column, beam and *dougong* (interlocking wood brackets), color painting decors, etc. Another is (at least some parts) the adoption of the western new architectural technology, such as ferroconcrete frame structure and truss. The combination and overlapping of the traditional Chinese style and the western architectural technology were first exercised by the occidental Christians and implemented mostly in religious related institutes, such as schools, hospitals, etc.

2.1 Christian Localization and Church Buildings

In the late Qing Dynasty and the early Republic of China, especially after the 1900 National Upheaval (or Boxer Uprising) and the May Fourth Movement, foreign western churches in China realized that adjusting their missionary course was necessary to cope with the worsening dilemma of the populace-religion conflict and the broadening China's nationalism awakening. In the early 20th century, finally, the Sinicization" of the Catholic and the "Localization Movement" in general undertaken by churches were settled as a norm, and they have been practiced ever since in China. In the 1920s, the "Committee of Coordination and Promotion of Christian Higher Education in China" sponsored directly by major (Protestant) Foreign Christian Missionary Societies[2] and Rockefeller Foundation of the United States proposed firmly to promote "efficient, Christian and Sinicized" Christian schools in China. Missionaries must not only dress Chinese robe, mandarin jacket, and skullcap, but also need to learn Chinese and study Confucian literature.[3] They embraced Chinese customs in diet and daily life and associated Christianity with Chinese populace in all aspects in order to eliminate the contravening reaction feuled by cultural differences. The architectural activities launched by churches became a major undertaking, by which churches and religious schools became the carrier of the two pioneering architecture styles. From the early 20th century to the end of 1920s, in this short two decades, there were 17 major religious schools built in

1 Fu Chaoqing (1993) *Contemporarily Styled Chinese Classical Architecture—Study on Bureaucratization History of Chinese New Architecture in 20th Century* (vii), SMC Publishing

2 Foreign Christian Missionary Society was a Christian missionary society with most members of Christian churches in western Europe and North America organized to engage in foreign missionary work, including founding churches, schools, newspapers and charities. See: Gu Changsheng (1991), *Misionaries and Modern China* (2nd ed., pp 109), Shanghai: Shanghai People's Publishing House

3 Gu Changsheng (1991), *Misionaries and Modern China* (2nd ed., pp 358), Shanghai: Shanghai People's Publishing House

succession, including Soochow University (Suzhou, 1902, by US), Aurora University (Shanghai, 1903, by France), Saint John's University (Shanghai, 1905, by US), Hangchow University (Hangzhou, 1910, by US), West China Union University (Chengdu, 1910, by US & UK), Huachung University (Wuhan, 1910, by US & UK), the University of Nanking (Nanjing, 1911, by US), Hwa Nan College (Fuzhou, 1914, by US), Hsiang-Ya Medical College (Changsha, 1914, by US), Ginling Women's College of Arts and Sciences (Nanjing, 1915, by US), Shanghai University (Shanghai, 1915, by US), Lingnan University Guangzhou, 1916, by US), Yenching University (Beijing, 1916, by US), Shantung Christian University (Jinan, 1917, by US & UK), Fukien Christian University (Fuzhou, 1918, by US & UK), Jingu University (1922, by France) and Fu Jen Catholic University (Beijing, 1929, by US). Among them, except Soochow University and Jingu University, the rest church universities were designed with the "architectural integration of Chinese and Western" strategy, in which, the contemporarily styled Chinese classical architecture, as proven to be more efficient, had definitely made its case with unique vantage.

Case 1 Holy Saviour's Cathedral

Holy Saviour's Cathedral is at today's 85 Tonglinge Lu of Xicheng District of Beijing (old name Nangouyan). It, also called Anglican cathedral, Holy Cathedral, and Nangouyan Saviour's Church, built in 1907, is the original primary church and the cathedral of the Anglican-Episcopal Church of North China Mission. It is an important testimony of the church architecture evolution in Beijing after 1900—retaining the church architectural format in layout yet built with Chinese construction materials of blue bricks, gray cylindrical tiles, roof, cornices and gables, and the architectural details in the hard-edged form of traditional northern China houses, while windows in span of brick-built arch, delivering a traditional Chinese architectural tone in the overall church appearance blended with the western classical church solemnness. It is a typical "architectural integration of Chinese and Western" building (Fig. 2-1).

With a flush gable roof for the two-story high main building sitting on a Latin cruciform plan, two lower transepts with a single-pitched roof each and two side portals in the center of each side of the south

Fig. 2-1 The Holy Saviour's Cathedral, Beijing

Fig. 2-2 Bird's eye view of the Holy Saviour's Cathedral, Beijing

section of the cloisters with their roof ridges close to the gable ridge of the main building, the roof of this church architecture forms two crosses in shape (Fig. 2-2), where an octagonal sunlight belvedere was built on the top of each intersection of the cross with the bigger one also used as a bell tower (Fig. 2-3). The main entrance of the church is at the south façade with a gable wall. Its two sides and the lintel have a white marble carving couplet set with the top one of the set hung on the lintel with an inscription "Enshrine and Worship" and the left one of the remaining pair "God's Temple" and the right one "Heaven's Gate". Right above the entrance, there is a circular rose window in Gothic Architecture. The church structure and its detailed craftsmanship are also good examples of the architectural integration of Chinese and Western style. Its mixed brick-and-wood bearing structure has 11

Fig. 2-4 Bottom view of the internal wood structure of the bell tower in the Holy Saviour's Cathedral, Beijing

Fig. 2-3 Exterior of the bell tower in the Holy Saviour's Cathedral, Beijing

Fig. 2-5 "Buttress" touch of the side-wing portals in the Holy Saviour's Cathedral, Beijing

Fig. 2-6 Campus of the University of Nanking (in the 1920s)

pin (a Chinese unit of frame) wood gable roof frames connected to the entablature (truss) of the isosceles triangle portion (pediment) of the gable. On the top the frame, there are square cross-section purlins and rafters covered with gray clay semicircle-shaped cylindrical tiles. The most amazing part is the octagonal triple eaves surrounding the sunlight belvedere, also the bell tower, above the altar and its octagonal base. The wood structure of eaves grows inward by stories, resembling the Hall of Prayer for Good Harvest of the Temple of Heaven (Fig. 2-4). The inner floor is of wood, so is the altar, surrounded by a set of altar fence in Chinese Annatto carved with anthemion and furnished with Chinese furniture. The church has a baptizing pool with a water inlet and drainage, which was rare at the time of its construction. The two sides of the main entrance and side-wing portals all have thick brick walls with a thickness of two and half bricks or 60 cm, suggesting a Gothic "buttress" touch (Fig. 2-5). Some of the window openings were built in semi-circular or chamber arch. Evidently, the spatial and bearing consideration hints the western approach. The decorations of the entrance, roof and internal arrangement are with Chinese taste. The details of the exterior wall design incline toward the western. The combination delivers an amazing space preference with a commingled cultural integration.

The church went through vicissitudes. It has witnessed China's great changes in the 20th century. Its original site was a private mansion of an official of the Ministry of Justice of the Qing Dynasty, Yin Keting. The SPG (Society for the Propagation of the Gospel in Foreign Parts) missionary Francis Lushington Norris took the advantage of the "1900 National Upheaval" in 1900 and plundered this place when the British Army besieged the Xuanwumen (or the Gate of Military Might) and its surrounding land of Beijing City. He forced Yin's son who was opposing his savage behavior to sign an unfair contract to sell the place worth tens thousands silver with only 8,000 silver *tael* (1 $^{1}/_{3}$ ounces) to the Anglican Church, followed by dismantling the mansion and turned it to Bishop Charles Perry Scott (1847—1927) of the Anglican-Episcopal Church of North China Mission, who hired a contractor to design and build the church in 1907. The church was put into use in the same year. It is the first church owned by the Anglican-Episcopal Church in China. In 1911, the competent authority of the Beiyang Government registered and updated the record to have Anglican Church officially become the property of the Anglican-Episcopal Church until 1949 when the church gradually evacuated from Chinese Mainland. In the following years, the church became the warehouse of the Television Technology Research Institute, and once it was in dilapidated state. At the end of the 20th century, a Hong Kong-owned enterprise Beijing Saiweng Information Consulting Services acquired the church building and invested 800,000 yuan in renovation. It had been using it as its office building since and until 2003 when it became a Beijing historical monument and cultural relic under state protection.

Case 2 Campus Buildings of University of Nanking

The University of Nanking, located at 22 Hankou Lu, Gulou District of Nanjing City, is now Nanjing

University. The University of Nanking, originally the Methodist Nanking University built in 1888, is the first church university established in Nanjing by the Methodist Church of the United States.

When the University of Nanking was registered in 1910, it purchased land in the southwest of Gulou District of Nanjing City. Architect Cady X. Crecory from New York did the design. US Architect Serverance and Surveyor Mohr of the architectural firm of Perkins, Fellows & Hamilton from Chicago were responsible for the construction. The campus architectural complex is the earliest successful case that exercised the complete process of thorough and professional planning and design of integration of Chinese palace architecture and the western construction system and technology (Fig. 2-6). Major buildings include:

Fig. 2-7 Auditorium in the University of Nanking

Fig. 2-8 Joseph Bailie Building in the University of Nanking

The Chapel, or the small auditorium, built in 1916, was designed jointly by Chinese architect Qi Zhaochang and the architectural firm of Perkins, Fellows & Hamilton of US;

The Church, today's auditorium, built in 1917, was designed by the architectural firm of Perkins, Fellows & Hamilton of US;

The Tower, today's North Building, built in 1919, was designed by US architect A. G. Small;

The Science Hall, today's East Building, built in 1925, was designed by US architect A. G. Small;

Joseph Bailie Building, today's West Building, built in 1926, was designed by Chinese architect Qi Zhaochang;

The Library, built in 1936, was design by Chinese architect Yang Tingbao.[1]

The construction duration of the above architecture complex spanned over 10 years yet still created a structurally and spatially harmonized campus environment, attributed to the foresight and sustainability of planning and designing efforts and awareness of its architects from the very beginning of the project. Theses buildings were built by phases. All walls were built with blue bricks, Chinese hip-and-gable roof covered with gray cylindrical tiles. The buildings' symmetrical layout was planned circumspectly with long depth and small window openings added with stone apron and lintel, delivering a composed and decorous effect of idiomorphism of Chinese palace architecture (Fig. 2-7, Fig. 2-8). The major manifestation of the "architectural integration of Chinese and Western" style includes Chinese architectural format and exterior decoration materials and western building system, structure and construction technology. Taking the North Building as an example, it occupies a construction area over 3,400 square meters and has two stories above the ground level and one basement level. The overall plan is rectangular. The structure is in brick-and-wood. Besides the traditional Chinese hip-and-gable roof, in the south façade the building has a five-story high square-shaped tower (Fig. 2-9), a western touch dividing the building into two sections, or the East and West wings. This was actually borrowed from the western classical space arrangement of a bell tower. It is surprising and impertinent yet the special cruciform ridge of the hip-and-gable roof and the single hip-and-gable roof of the two symmetrical side wings added to the top and the dazzling red five-star placed at the tip of the tower after 1949 "orchestrate" a "nationalistic symphony" with apropos finish, elementally rich and eccentrically melodic and acoustic. Compared with the library later designed by Chinese architect Yang Tingbao returning from his US study endeavor, the profound design with Chinese official architecture style apparently overwhelmed the US trained architect from Chicago. From another perspective, this also reveals the merit of having a sole construction contractor for the harmonization of architectural styles of these buildings designed by different architects at different historical periods. This construction contractor was the reputed Chen Ming Kee Construction.

Fig. 2-9 North Building in the University of Nanking

The architectural complex of the University of Nanking is preserved in sound condition. For its significance, in 1991, it was selected as the outstanding

1 Lu Haiming, Yang Xinhua, & Pu Xiaonan (2001), Nanjing Architecture of the Republic (pp 158—168), Nanjing: Nanjing University Press

Fig. 2-10 Campus of Ginling Women's College of Arts and Sciences

Fig. 2-11 A large lawn at the main entrance to Ginling Women's College of Arts and Sciences

building of contemporary China. In 2006, it was designated as a historical monument and cultural relic under state protection nominally under the relics of the "the Methodist Nanking University".

Case 3 Campus Buildings of Ginling Women's College of Arts and Sciences

Ginling Women's College of Arts and Sciences, also known as Ginling Women's University before 1930, located at 122 Ninghai Lu, Gulou District of Nanjing City, is now Nanjing Normal University.

In 1913, American churches of the United States, including the Presbyterian Church in the United States of America, the Methodist Episcopal Church,

Fig. 2-12 Side view of the details of the hip-and gable roof in a teaching building of Ginling Women's College of Arts and Sciences

the Methodist Episcopal Church South, the American Baptist Churches and the Christian Church formed an alliance and decided to establish a women's university in Yangtze River Basin. On November 13 of the same year, they organized a university council and decided the school site. In 1915, Ginling Women's College of Arts and Sciences was open at the old private garden site of Li Hongzhang in Nanjing. The first term of the college president was Mrs. Laurence Thurston. In July 1923, the college was moved to Sui Yuan (the private garden of Yuan Mei). The council invited American architect Henry K. Murphy to preside over the planning and design of the campus, followed by later participations of the well-known Chinese architect Lü Yanzhi (see *Encyclopedia of Modern China*: pp 60, 2009 Charles Scribner's Sons; Britannica: *Chinese Architecture*, by Liu Qiyi) and Chen Ming Kee Construction for the construction project, which was commenced in 1922 and completed in 1923. The campus complex includes an assembly hall, science building, literature building and four student dormitories, totaling seven palatial buildings. In 1934, to complete the Chinese palace campus complex, the library and auditorium were added. It has earned its reputation of "The Most Beautiful Campus of the East" (Fig. 2-10, Fig. 2-11).

The campus was planned symmetrically over an east-west axis. The front entrance leads to an avenue

Fig. 2-13 Details of windows on the gable wall under the hip-and gable roof in a teaching building of the Ginling Women's College of Arts and Sciences

Fig. 2-14 Original colour design of the traditional eaves in Ginling Women's College of Arts and Sciences

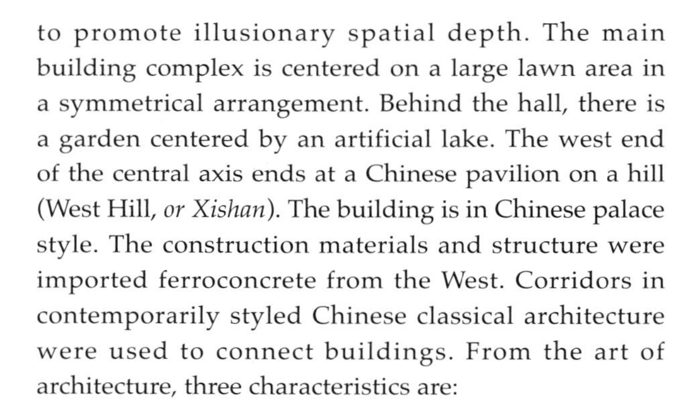

to promote illusionary spatial depth. The main building complex is centered on a large lawn area in a symmetrical arrangement. Behind the hall, there is a garden centered by an artificial lake. The west end of the central axis ends at a Chinese pavilion on a hill (West Hill, *or Xishan*). The building is in Chinese palace style. The construction materials and structure were imported ferroconcrete from the West. Corridors in contemporarily styled Chinese classical architecture were used to connect buildings. From the art of architecture, three characteristics are:

1. It represents the overall structure of Chinese official architecture, including the ratio, dimensions and details of the roof, such as fascia boards of the hip-and-gable wall and gable ridge with drippers, in vivacious fashion (Fig. 2-12). As a crucial point, the long being criticized misalignment of *dougong*, chapiter and ambiguous structural logic is proven to be irrelevant to the nonprofessional users and audiences of the architecture, who are not concerning with the structure rationale, but the application of *dougong* and the appearance of *dougong*, instead.

2. On the approach aspect, using windows to emphasize the high spatial efficiency with reasonable utilization of the internal space with the large sloping roof structure is a way to improve the usable area coefficient of the design. This is a popular method adopted in the western classical architecture for folk dwellings, while new to Chinese official architecture. In this case, the design by adding a window on the gable wall (pediment) to input natural sunlight into the gable roof is to match more reasonably the internal space utilization—to be used as a loft storage room or small office. This approach brings a new dimension for the walls of Chinese hip-and-gable style (Fig. 2-13).

3. It pioneered the abstract expression for Chinese official architecture. From the design aspect, adopting directly the color painting of Qing's official architecture is the easiest as it was used for the former National Library of Peiping (today's National Library of China). However, architect Henry K. Murphy, in this case, put forward tremendous efforts in designing rectangular patterns for beams, architraves, and chapiters and even adding greasepaint in dark green, olive green, light green, white, black and bright red, while for *dougong*, adding highly bright light green and light blue (Fig. 2-14). Although, from a short-distance perspective view, it appears inevitably blunt, however, by stretching the view farther, the appreciation of the traditional color painting cannot be neglected. Many details on eaves, while under shade, for the high

Fig. 2-15 A portrait of Lü Yanzhi (1894—1929)

brightness, do not look tepid, tarnished and sluggish.

In general, with the impetus of "Sinicization" policy of church activities, before 1920s, the church buildings and the campus building of church schools were designed with assertive localization efforts, and the motivation was to adapt. Attempts undertaken during this stage were initiated by missionaries and western architects, leaving "seams" and "traces" of cohesion undisguised due to insufficient theoretical background of Chinese wood architecture. However, The integration of the western modern construction system and technology with the traditional Chinese architecture format and delicate craftsmanship was unprecedented. Many "ingenuities" presented in their efforts were meritorious. A noticeable attainment

Fig. 2-16 Nanjing Sun Yat-sen's Mausoleum

of this groundbreaking pragmatic process was the cultivation of the first generation Chinese architect genus—Lü Yanzhi, whose works of Nanjing Sun Yat-sen's Mausoleum and Guangzhou Sun Yat-sen Memorial Hall are two epical masterpieces of the "Contemporarily Styled Chinese Classical Architecture".

2.2 Experiments of Ingenuity: Nanjing Sun Yat-sen's Mausoleum and Guangzhou Sun Yat-sen Memorial Hall

Starting in late 1910, Chinese scholars studied architecture in Japan, Europe and US returned to their homeland to begin their career, so did the first group of contemporary Chinese architects step onto the stage of history. This changed the millennial old tradition of impartment by mouth and heart and also facilitated the ascendance of Chinese architectural design market dominated by western architects, a situation due to non-professionalism in Chinese civil engineering and architect occupation. This installed the historical recognition of the first generation of Chinese architects. Among them, Lü Yanzhi (1894—1929), graduated from the Department of Architecture of Cornell University and worked with Henry K. Murphy, was the extraordinary one (Fig. 2-15). With only 36 years, He created a glorious chapter for the "Contemporarily Styled Chinese Classical Architecture"—Nanjing Sun Yat-sen's Mausoleum and Guangzhou Sun Yat-sen Memorial Hall.

Case 1 Nanjing Sun Yat-sen's Mausoleum

Nanjing Sun Yat-sen's Mausoleum, situated at the foot of the second peak of Tzu-chin Mountain (or ZiJinshan) in the east suburban of Nanjing, China, adjacent to the Xiaoling Mausoleum of Emperor Taizu of Ming in the west and the Linggu Temple in the east, is the tomb and the memorial building cluster of the forerunner of China's democratic revolution Dr. Sun Yat-sen (1866—1925). The construction of the Mausoleum was started in January 1926 and the burial ceremony of Dr. Sun Yat-sen was performed on June 1, 1929. In 1961, it was designated as a historical

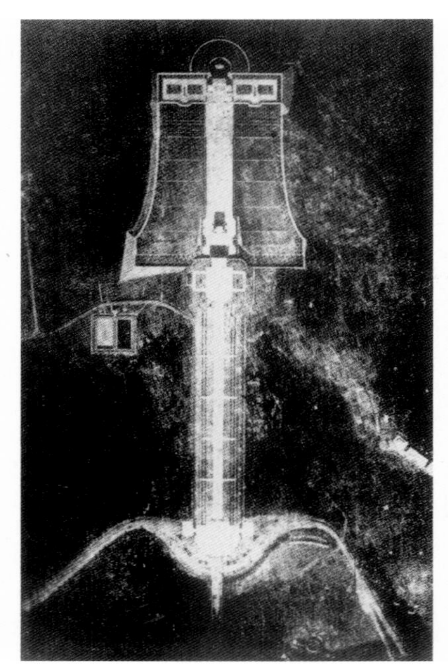

Fig. 2-17 An aerial photo of the preliminarily built Nanjing Sun Yat-sen's Mausoleum

Fig. 2-18 Buildings of Nanjing Sun Yat-sen's Mausoleum rigorously aligned along the central axis

Fig. 2-19 Memorial Arch of Love for All Mankind in Nanjing Sun Yat-sen's Mausoleum

monument and cultural relic under state protection (Fig. 2-16).

On March 12, 1925, Dr. Sun Yat-sen passed away in Beijing. On May 13 of the same year, the "Prime Minister Funeral Preparatory Committee" decided to collect mausoleum design proposals with reward from home and abroad. The 32-year old, little known Lü Yanzhi applied. After studying carefully imperial mausoleums of ancient China and Europe, he elaborately designed the tomb with great originality along with detailed design description, including the preliminary overall layout, materials and color. Judged by consultants Ling Hongxun, Emil Busch (a famous German architect), Wang Yiting, Li Jinfa in written opinion, Lü was awarded the first prize, followed by the second place Fan Wenzhao, the third place Yang Xizong, and seven other honoraries. The "Prime Minister Funeral Preparatory Committee" presented all of designs of the awarded candidates to the public in Shanghai for five days to request for suggestions. During the exhibition, visitor swarmed the exhibition and news media of home and abroad published critics and interviews, causing a sensation in the Bund. After the exhibition, the Committee reevaluated the designs and unanimously agreed that Lü's design was "Simple yet elegant, pertinent to the quintessence of the ancient Chinese architecture". The decision on Lü's design was finalized and Lü was hired as the chief architect to preside over the construction drawings, materials, construction oversight and final project acceptance. Shanghai Yao Xin Kee Construction won the construction tender. Since the commencement in January 1926, the construction project had been going through turmoil of the Warlord Era and the Northern Expedition in China. The main construction was completed finally in May 1929. On June 1, Dr. Sun Yat-sen's burial ceremony at the Mausoleum was held. Lü, unfortunately, died of cancer in Shanghai on March 18, 1929 due to the accumulated overwork of the construction design and supervision of the

Fig. 2-20 Portal of Nanjing Sun Yat-sen's Mausoleum

Fig. 2-21 Tablet pavilion of Nanjing Sun Yat-sen's Mausoleum

Fig. 2-22 Stone lions at the portal of Nanjing Sun Yat-sen's Mausoleum

Fig. 2-23 Bottom view of a symmetrical pair of copper dings (cauldrons) in front of the Sacrificial Hall of Nanjing Sun Yat-sen's Mausoleum

Fig. 2-24 Symmetrical copper dings and ornamental columns in front of the Sacrificial Hall of Nanjing Sun Yat-sen's Mausoleum

Fig. 2-25 The Sacrificial Hall of Nanjing Sun Yat-sen's Mausoleum

Fig. 2-26 Towers at corners of the Sacrificial Hall of Nanjing Sun Yat-sen's Mausoleum

project. Nanjing National Government gave him commendation.

The main constructions of the Mausoleum include the memorial arch (or *paifang*), access, portal, tablet pavilion, sacrificial hall and marble sarcophagus chamber. In addition, a series of commemorative buildings, such as the Yongmu Cottage, cauldron (or *baoding*), bandstand, Liuhui Terrace Pavilion,

Fig. 2-27 Exquisitely executed connection between the hip-and-gable roof with two layers of eaves and a corner tower of the Sacrificial Hall of Nanjing Sun Yat-sen's Mausoleum

Reverence Pavilion, Gloriole Pavilion, Devotion Pavilion and Dr. Sun Yat-sen Museum were added. From the aerial view, the entire layout of the Mausoleum resembles a "Liberty Bell" (Fig. 2-17). For years, it is said the bell symbolizes the message "revolution is not finished" and Dr. Sun's will that "must arouse the people to continue the course" should prevail. The influence like other extraordinary arts has gone beyond the original intention of the creator. To Lü, this holds the same.

The design art of the Mausoleum conveys five features:

1. It has elaborated layout and structure. The site design follows the landscape closely. The standalone buildings of the memorial arch, access, portal, tablet pavilion, sacrificial hall and marble sarcophagus chamber are aligned on the same central axis and connected by large patches of gardens and stone stairs into a cluster, diffusing awe-inspiring ambiance (Fig. 2-18).

2. The grafting of Chinese and Western is in a perfect integration. The marble sarcophagus chamber is behind the sacrificial hall, adhering to the traditional Chinese rites. With western stone architecture features, the memorial hall has the base tone of the official architecture of the Qing Dynasty. The marble sarcophagus chamber is entirely in western architecture. The overall integration is seamless. The ancient traditional Chinese tomb characteristics are preserved on memorial arch, portal and tablet pavilion (Fig. 2-19, Fig. 2-20 and Fig. 2-21), and the ornamental columns, stone lions and copper ding (cauldron) (Fig. 2-22, Fig. 2-23 and Fig. 2-24) are added to emphasize Chinese inherent characteristics with the embellishment of western cultural attributes, uniquely identifying the originality of the design. For example, the sacrificial hall has a hip-and-gable roof with two layers of eaves resembling the official architecture of the Qing, yet with modification—the architecture under the eaves mount is the western classical square fort structure, upstanding and concrete. Four towers are protruding at four corners of the fort structure. The whole building emanates western stone architecture

Fig. 2-28 Bottom view of the Sacrificial Hall of Nanjing Sun Yat-sen's Mausoleum from the stone stairs

Fig. 2-29 Overhead view of stone stairs from the platform of the Sacrificial Hall of Nanjing Sun Yat-sen's Mausoleum

Fig. 2-30 Roof details of the Sacrificial Hall of Nanjing Sun Yat-sen's Mausoleum: abstract and simplified deity and beast statuettes

Fig. 2-31 Cornice details of the Sacrificial Hall of Nanjing Sun Yat-sen's Mausoleum: imitated wood stone pillar, beam, architrave and tangent circle pattern

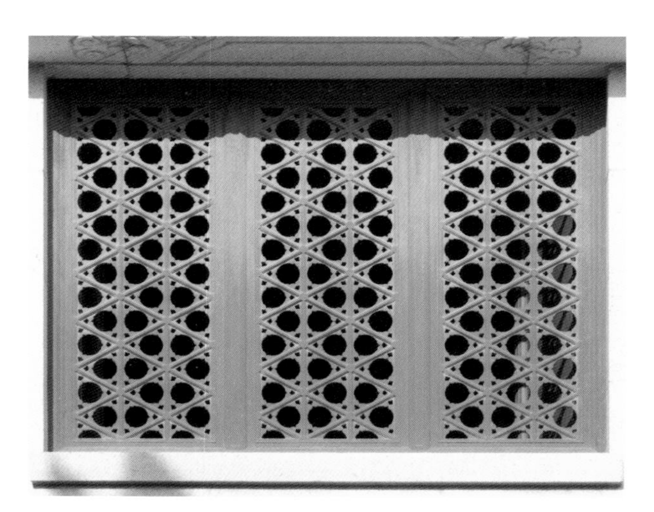

Fig. 2-32 External wall details of the Sacrificial Hall of Nanjing Sun Yat-sen's Mausoleum: imitated wood window cast with purple copper

Fig. 2-33 Exterior of Guangzhou Sun Yat-sen Memorial Hall

temperament under a Chinese official architectural roof (Fig. 2-25, Fig. 2-26 and Fig. 2-27).

3. It features an intriguing perspective design. The mausoleum is built on a hill with 392 stone steps interconnected by eight terraces, which are arranged with perspective variations by steeper slope upward (the elevation angle is 9° viewed from the Memorial Arch of Love for All Mankind at the Entrance Gate to the sacrificial hall and 19° from the tablet pavilion), so that when climbing upward, the arrangement delivers a solemn feeling and when looking downward from the top terrace, only the platform formed by the connected terraces and no stairs is in the view (Fig. 2-28). It is a quite marvelous design (Fig. 2-29).

4. It has properly designed details and color paintings. It makes proper abstraction of deity and animal patterns used in Qing's genre (Fig. 2-30, Fig. 2-31) with matching coloring. The sacrificial hall was originally designed to use copper roof tile but switched to azure-glazed roof tiles the same as the memorial arch and the tablet pavilion because of cost and theft consideration, providing a consistent coloring scheme as the traditional Chinese imperial tombs.

5. The architecture is inspiring. The overall layout and attributes induce musing. The plan forms a large bell pattern, mimicking an alarm prevising the world (from Ling Hongxun). "forming a big bell shape, an amusing articulation" (from Li Jinfa) and "resembling his mettle and spirit" (from Wang Yiting). The "intention" is simple, while the "indication" is boundless, so is the inspiring creativity.

The design and construction of Nanjing Sun Yat-sen's Mausoleum reflect Lü's architectural trait in integrity—adopting the western construction technology to implement the "China's inherent architectural style". The access, portal, tablet pavilion, sacrificial hall and stairs and the buildings underground and above the ground level of the marble sarcophagus are all in ferroconcrete. "Wood elements", such as *dougong*, beam and columns, façade, tablet on the memorial arch, used to be in wood, were built with ferroconcrete also. The "imitating wood" doors of the portal and the sacrificial hall were cast with purple copper (Fig. 2-32). The successful application of these construction technologies is because the western modern construction skills and materials coalesced seamlessly with the Chinese official architecture.

The Mausoleum was designed as the first large project of "contemporarily styled Chinese classical architecture" by the Chinese architect. Its success is significant in demonstrating the use of historical elements in new architectural design. Betokened by it, the "architectural integration of Chinese and Western" sets the trend and becomes a genre of the China's contemporary architecture, as commented once by the well-known Chinese architect Liang Sicheng, "The Mausoleum although weighing the western attribute sure is an exemplar for new designs to recount the ancient Chinese designs, and a milestone of China's renaissance.[1]

Case 2 Guangzhou Sun Yat-sen Memorial Hall

Guangzhou Sun Yat-sen Memorial Hall, situated at 259 Dongfeng Middle Lu, Yuexiu District, Guangzhou, occupying over 12,000 square meters, 57m in height, is a masterpiece of contemporary architecture of Guangzhou. It is also a historical monument and cultural relic under state protection (Fig. 2-33). The construction work started in January 1929 and completed in November 1931.

Not before long the death of Dr. Sun Yat-sen, the construction of the Memorial Hall was under planning. The Guangzhou Sun Yat-sen Memorial Hall Preparation Commission was not formed until the early 1927, chaired by Guangdong Provincial Government Chairman Li Jishen. In April 1927, the Preparation Commission published on news media of home and abroad to solicit design for the Memorial Hall and Dr. Sun Yat-sen Monument. Lü Yanzhi, the architect already being hired to preside over Nanjing Sun Yat-sen's Mausoleum responded to the solicitation by submitting the design drawing prepared overnight. In the mid-May of the same year, Lü's design was awarded the first prize and he, two other structural engineers who were also Lü's alumna of the same US alma mater, Li Keng, Feng Baoling, architects Lee Gum Poy, Qiu Xiejun and Ge Hongfu were hired to proceed on the construction project. In November 1927, Shanghai Voh Kee Construction won the construction tender at a bidding price of 928,985 silver *tael* (accounting currency). In April 1928, the complete drawing set was finished. The construction started on April 23. Delayed by funding, the project was not finished until October 10, 1931.

Fig. 2-34 Bottom view of Guangzhou Sun Yat-sen Memorial Hall (from Zhao Yunfei, South China University of Technology)

Fig. 2-35 Bird's eye view of Guangzhou Sun Yat-sen Memorial Hall from Yuexiu Mountain

1 Liang Sicheng (1998, 354). *Chinese Architecture History*, Tianjin: Baihua Literature and Art Publishing House

Fig. 2-36 Vividly imitated wood structure of the exterior of Guangzhou Sun Yat-sen Memorial Hall (from Zhao Yunfei, South China University of Technology)

Fig. 2-37 Details of a façade of Guangzhou Sun Yat-sen Memorial Hall (from Zhao Yunfei, South China University of Technology)

Guangzhou Sun Yat-sen Memorial Hall is the same as Nanjing Sun Yat-sen's Mausoleum in "Chinese palace" architecture. The roof is covered with azure-glazed roof tiles. The building has a height of 57 meters and a construction area of 8,700 square meters. The auditorium accommodates more than 4,700 seats, where the balcony has 2,200 seats.[1] Its entire octagonal pyramidal roof is supported by using cantilever steel trusses with 30m span and four 45° interconnected triangular trusses. It was the auditorium with the longest span in China then. Dr. Sun Yat-sen Monument

1 Yang Yongsheng, & Gu Mengchao (Ed.)(1999), *Chinese Architecture of 20th Century* (pp 134), Tianjing: Tianjin Sience & Technology Press

(or Dr. Sun Yat-sen Tower) has a height in 37 meters in 12 stories (Fig. 2-34). A stair from the ground floor of the Monument leads to the top floor, giving a magnificent overlook view of the Pearl River (Fig. 2-35). The Tower and the Memorial Hall are connected by over 300 stone steps delivering a regal sensation. It is praised as the most renowned building in Lingnan (referring to lands in the south of China's "Five Ranges" which are Tayu, Qitian, Dupang, Mengzhu, Yuecheng).

Compared with the Mausoleum in Nanjing, the “Sinicization” is extended in the art of architecture and more efficient in structural technology—a leap toward more “Contemporarily Styled Chinese Classical Architecture”. As to the appearances, excluding the glazed roof top, the sacrificial hall and tablet pavilion of the Mausoleum as upstanding and concrete variants, incline toward the western classical architecture. The Memorial Hall, adding more touches to the exterior wall with imitated wood columns, beams and architraves, including the color painting, brings it closer to Chinese official architecture, or, it has captured the essence of the wood architecture—the open space between colonnades and the rhythm that they emphasize (Fig. 2-36, Fig. 2-37).

In addition, the octagonal pyramidal roof uses 30-meter span Fink steel roof truss, which creates the record of the largest spanning space in Chinese contemporary assembly halls. It is an exemplar of

Fig. 2-38 Exterior of the octagonal conical roof with two layers of eaves in Guangzhou Sun Yat-sen Memorial Hall

Fig. 2-39 Interior of the octagonal conical roof with two layers of eaves in Guangzhou Sun Yat-sen Memorial Hall

Fig. 2-40 Octagonal conical roof with two layers of eaves and spectator seats in the balcony of Guangzhou Sun Yat-sen Memorial Hall (from Zhao Yunfei, South China University of Technology)

Fig. 2-41 Steel roof truss of the octagonal conical roof with two layers of eaves in Guangzhou Sun Yat-sen Memorial Hall

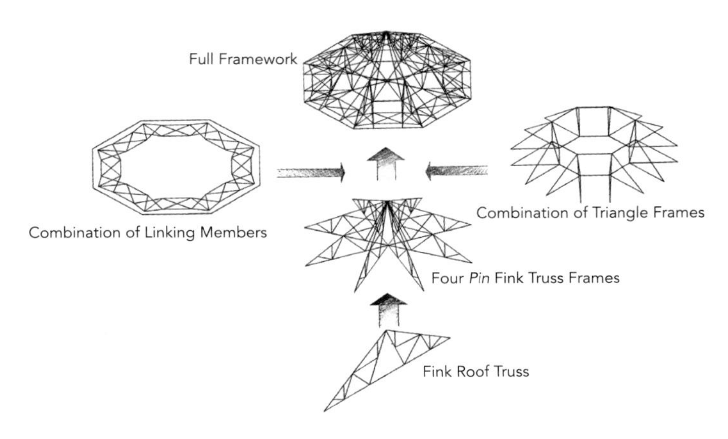

Fig. 2-42 A group of analysis graphics showing the steel roof truss of the octagonal conical roof with two layers of eaves in Guangzhou Sun Yat-sen Memorial Hall

using truss for a complicated roof in "China's inherent architectural style" (Fig. 2-38, Fig. 2-39 and Fig.2-40). Its merits include:

1. Four *pin* gable frames (Fink steel truss) are put over the shear wall structure attached to the eight-side inward bending polygonal trusses above the octagonal auditorium hall that support the octagonal berm.

2. Four *pin* Fink steel truss frames intersect symmetrically at the center of the octagonal roof vertex. Each *pin* truss frame contains smaller scalene triangular truss frames with all their longer side of the acute triangle connected to the tip of the octagonal roof vertex and added with purlin to form the main structural skeleton of every triangular roof gable frame. Between the bottom sides of two acute triangles and the bottom of the Fink steel truss, shears are added to enforce the overall integrity of the steel structure. The final finish is a complicated structural system of spatial building frame (Fig. 2-41, Fig. 2-42).

Compared to the ferroconcrete truss structure used for the roof of the Mausoleum, the steel truss frame is lighter, structure-wise more efficient and easier to construct. This was attributed to its chief architect Lü's wise choice and the joint contribution of the two structural engineers Li Keng and Feng Baoling. For their adding more "Sinicization" attributes and adopting more efficient structural technology of architecture, the design idea of the "Contemporarily Styled Chinese Classical Architecture" is elaborated further. It also defines the historical pioneering position of Lü Yanzhi in the "architectural integration of Chinese and Western".

2.3 Nanjing National Government and "China's Inherent Architecture"

Nanjing, as the political center of the National Government after 1927, had a group of functional edifices in "China's inherent architectural style" built within ten years, including commemorative buildings (Tan Yankai Tomb, Nanjing National Revolutionary Army Memorial Cemetery, Dr. Sun Yat-sen Museum), party and government office buildings (the KMT Central Supervisory Commission, Ministry of Railways, Ministry of Communications, and Examination Yuan), scientific research, culture and education buildings (the Institute of History and Philology, Institute of Geology and Institute of Social Sciences, Academia Sinica; Library of Ginling College), museums (the National Central Museum, Historical Museum of KMT's Central Party History), sports and recreational buildings (the Central Stadium, Inspirational Community Center), and hotels and official mansions (Hotel of Overseas Chinese, Chiang Kai-shek Villa), which provoked an architectural idiosyncrasy deeply related to the *New Chinese National Capital Planning*.

On April 18, 1927, Chiang Kai-shek declared officially the establishment of "Nanjing National Government". To beef up the city construction and administration, Nanjing National Government formed the Public Works Department in July 1927, the "New Chinese National Capital Construction Committee" in January 1928, and the "New Chinese National Capital Design Technical Specialist Office" on February 1. In December 1929, the *New Chinese National Capital Planning* was announced. At that time, the enthusiasm of overseas intelligentsia turning focus on the traditional Chinese culture was at its extremum attributed to two reasons: One was the unprecedented upsurge of nationalism to establish an independent sovereign democratic republic. Another was the consummate introspection of the intelligentsia's attitude from admiring the western civilization to uncertainty of its adaptability to China after the World War I, while searching for opportunities from the traditional Chinese culture to redeem national esteem, confidence, assertiveness and mettle. Driven by political "nationalism" and intelligentsia's "cultural quintessence renaissance advocacy", the National Government in its *New Chinese National Capital Planning* prescribed specific provisions of "China's inherent architectural style" as the official architecture for "governmental offices and public buildings"[1]. These policies set forth the orientation of architectural designs in Nanjing, Shanghai, Guangzhou and other large- and mid-size cities and stimulated the culmination of the "Contemporarily Styled Chinese classical architecture" until the retreat of Kuomintang regime to Taiwan.

Case 1 National Central Museum

The National Central Museum, the first modern museum of China, is today's Nanjing Museum (referred to as the "Museum") situated at 321 Zhongshan East Lu, Nanjing (Fig. 2-43). As most designs of the "Contemporarily Styled Chinese Classical Architecture" were blueprinted by the style of Qing's palatial edifices, the Museum was surprisingly designed in architectural style of the Liao Dynasty, primitive and solemn, departed from other official buildings not only in style but also in quaint construction process.

In 1933, Chinese intelligentsia led by Cai Yuanpei, jointly proposed to build a national museum. Headed by Historian Fu Sinian, the Museum preparatory department was organized in April of the same year. Geologist Weng Wenhao, anthropologist Li Ji,

Fig. 2-43 The main hall of the National Central Museum, Nanjing

and machinery and mining and metallurgy scientist Zhou Ren were in charge of the preparatory work of "Nature", "Humanity" and "Craft" halls, respectively. The Museum's "Construction Commission" was formed in July of the same year, chaired over by Weng Wenhao with Zhang Daofan, Fu Rulin, Fu Sinian, Ding Wenjiang, Li Shuhua, Liang Sicheng , Lei Zhen and Li Ji as commission members. Zhang Daofan, Fu Sinian and Ding Wenjiang served standing committee members and Liang Sicheng was appointed as a senior member. The primary function of the "Construction Commission" was to manage the funding, select the construction site, determine the construction plan, and supervise the construction work, etc. In November 1935, the Ministry of Education and the Academia Sinica jointly organized the Museum Trustee Council, which was headed by Cai Yuanpei. The "Construction Commission" and the Museum Trustee Council coordinated to secure the funding, site and land acquisition. The funding came from two sources: One was the appropriation of 1.5 million silver dollars allocated by the "Managing Council of Sino-British Boxer Indemnity Restitution" and another was the subsidy from Academia Sinica. In April 1935, the site of 100 *mu* (a Chinese land unit, 15 *mu* = 1 hectare) at the former royal fields of Banshan Garden was approved by Nanjing National Government. The site, facing Zhongshan Lu in the south, adjacent to Laoqi

1 *New Chinese National Capital Planning* (1929), New Chinese National Capital Design Technical Specialist Office (pp 35), New Chinese National Capital Design Technical Specialist Office

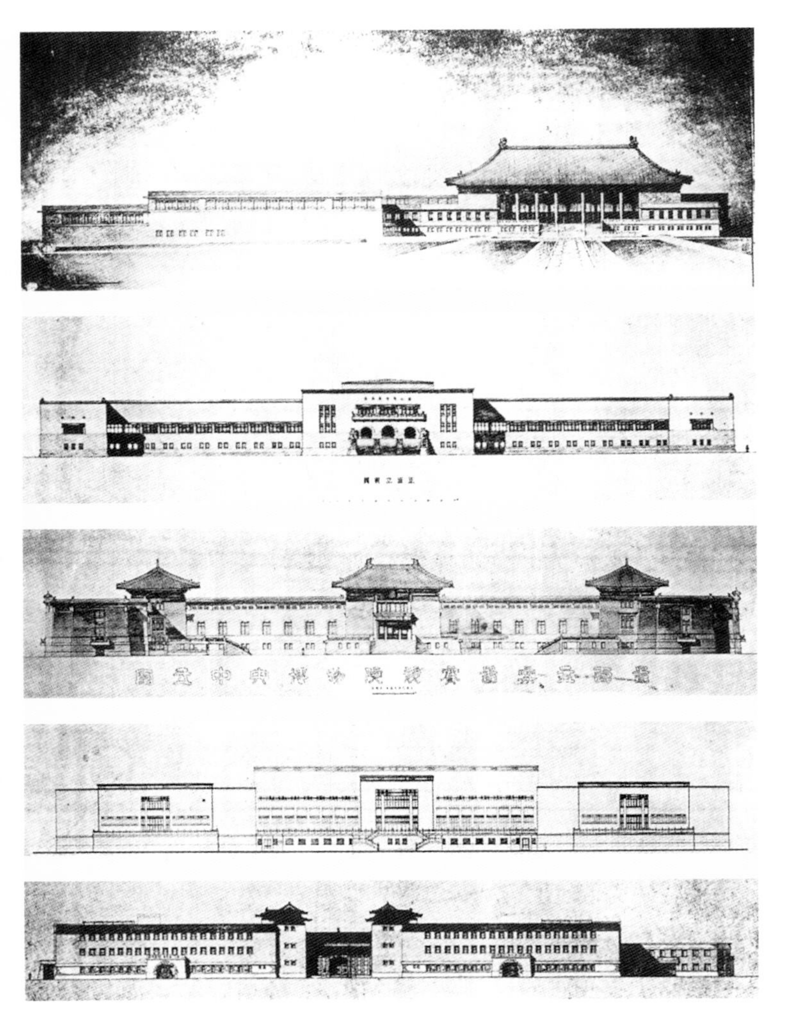

Fig. 2-44 to Fig. 2-48 Comparison by main façade of prize-winning works from the National Central Museum design competition (Design plan of Xu Jingzhi, Lu Qianshou, Yang Tingbao, Xi Fuquan, and Tong Jun from top to bottom according to the rankings)

Street in the east and close to the city artery leading to the main railway station under planning in the west, was 468m long from the south to the north, 173 meters at the maximum span from the east to the west. Its southwest corner was the rectangular-shaped "National Revolutionary Bereaved Children School Garden". Therefore, the land parcel of the site was "chopper-shaped", missing the southwest corner. The land slopes by six meters in elevation from the southeast to the northwest. The "chopper-shaped" site imposed a challenge to architects[1].

Soon afterwards, Liang Sicheng personally drafted the *National Central Museum Construction Commission Solicitation of Architecture Design Guidelines* in which he prescribed the pre-requisite of a coherent "Chinese architecture of technically maximized usability of the total construction area of 275,000 square feet (about 25,548 m^2), not compromising the requirement of a modern museum". In the *Guidelines*, he also called out the design tendering would not be an open bid but by invitation of "thirteen architects of Chinese citizen". The invited candidates were Li Zongkan, Lee Gum Poy, Xu Jingzhi, Xi Fuquan, Zhuang Jun, Chen Rongzhi, Lu Qianshou, Tong Jun, Guo Yuanxi, Dong Dayou, Yu Binglie, Yang Tingbao and Su Xiaxuan. Among candidates, except Su Xiaxuan politely declined the invitation, the rest all submitted their designs. After the five-member "Architectural Design Review Panel", Zhang Daofan, Hang Liwu, Liang Sicheng, Liu Dunzhen and Li Ji, carefully reviewed the designs, the anonymous vote favored Xu Jingzhi, followed by Lu Qianshou, Yang Tingbao, Xi Fuquan and Tong Jun in order of the vote[2] (Fig. 2-44 to Fig. 2-48). The three of the five winning designs were in "China's inherent architectural style", only Yang's design was in Liao-style (the Liao Dynasty style) architecture. As the director of architectonic department of the Society for the Study of Chinese Architecture, Liang Sicheng found the oldest wood architecture, Guanyin Pavilion of the Dule Temple in Jixian of Tianjin City built in the Liao Dynasty before his discovery of the Main Hall of the Foguang Temple. As a result, Liang instructed Xu to change his Qing's imitating style to Liao's, and Xu, as advised, finished his first phase construction drawings.

1 Li Haiqing, & Liu Jun (2001), *Growing through Exploration—Untold Story on Problems of Old National Central Museum Architecture*, Huazhong Architecture (Issue 6, pp86)

2 Li Haiqing, & Liu Jun (2001), *Growing through Exploration—Untold Story on Problems of Old National Central Museum Architecture*, Huazhong Architecture (Issue 6, pp86)

In April 1936, Jiang Yu Kee Construction won the construction bid. The construction project of the Museum started in June but interrupted by the Anti-Japanese War in August 1937 after three fourths of the first phase construction project was done with only six months work remaining. The interruption lasted eight years. During the war, the Japanese army converted the constructed portion of the Museum into an air defense command center. The renovations damaged many sections. After the war, the Commission wanted Jiang Yu Kee Construction to continue the construction, but another construction firm, Shanghai Construction Factory of Lugen Kee Construction (owned by Lu Genquan) with mob and political connections intervened in the project through the Bureau of Investigation and Statistics Director-General Zheng Jiemin. Meanwhile, the National Government embroiled in the civil war was not able to provide funding on schedule, while Lu Genquan was occupied with the military defense constructions for Shanghai. In 1947, the Museum had finally recovered from its pant to have the overall skeleton finished, but not until the early 1950s, completed the "Humanity" hall, the National Central Museum, or the old Nanjing Museum of today.

Xu emphasized on using symmetrical axis and pervasive spatial layers in perspective and landscape. The main building is situated away from the arterial Zhongshan Donglu, with reserved lawn patches, square and green zones. The hall has three layered terraces. The design brings out the grandeur of the main structure. Imitated in Liao's style, the bare structure and dimensions follow *Yingzao Fashi* (the *Treatise on Architectural Methods* or *State Building Standards*, a technical treatise on architecture and craftsmanship), while the details and decors combine the style legacy of the Tang and Song dynasties (Fig. 2-49). The hall is built over an area of seven bays in dimensions, with the roof styled in four-hipped-roof (or, *Si'ading*) and covered by brown glazed roof tiles. The design of the exhibition hall adopting modern museum approach has a flat roof added with oversail eaves cantilevered over its exterior walls to be in consort with the main hall design.

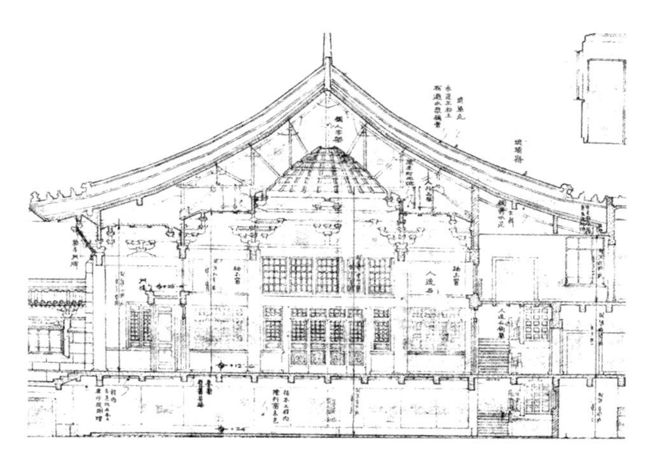

Fig. 2-49 Profile of the main hall of the National Central Museum (construction drawing)

Compared with the original awarded designs (*also see* Fig. 2-44 to Fig. 2-48), from the first place to the fifth, the enthralling points revealed are: 1. the favoritism of the "traditional Chinese curved roof"; 2. the favoritism of large "traditional Chinese curved roof"; 3. From the advise of switching from the imitated Qing style to that of the Liao given to Xu by Liang, the adoration of the antiquated "traditional Chinese curved roof". The immediate cognition induced by the intuitional affection of the symbolism of the traditional Chinese architecture was so strong that its explosion in the particular historical era had supplanted the "modern architecture". Xi Fuquan's design ranked the fourth place had its chance to avail. After the Anti-Japanese War, Xi extended his aspiration with the National Great Hall of the People, Nanjing (today's Nanjing Great Hall of the People) project. The outcome was not surprising. "China's inherent architecture style" was proven to be not only an impression deeply rooted in the minds of intelligentsia but also an instrument of political leverage worth to pursuit. It also unveiled conspicuously the amplified expressive function in the general aesthetic perception. Changed from the original imitation of Qing's style to Liao's style by following Liang's advice, the finalized implementation by Xu Jingzhi also indicated the nationalism ideology of incorporating into "China's inherent architectural style", any new archaeological discoveries conducive

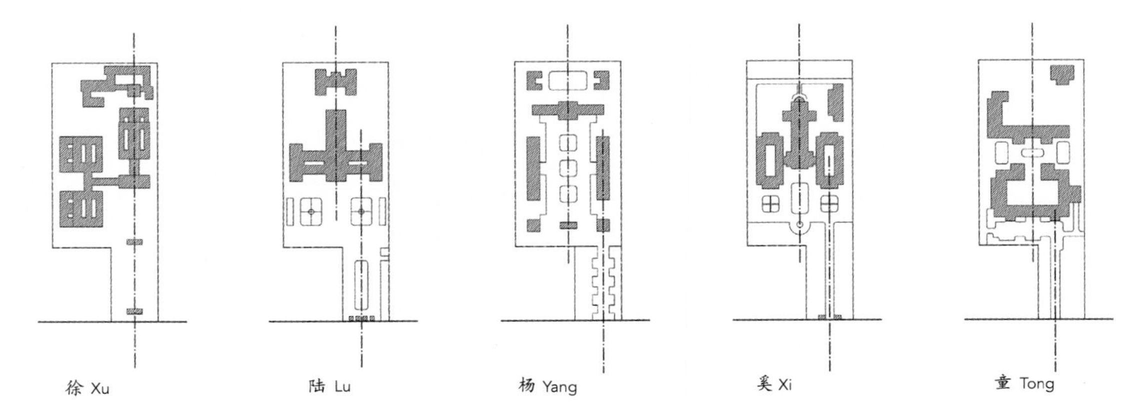

Fig. 2-50 Comparison by general view of prize-winning works from the National Central Museum design competition (Design plan of Xu Jingzhi, Lu Qianshou, Yang Tingbao, Xi Fuquan and Tong Jun from left to right according the rankings)

to re-glorify the long and brilliant Chinese civilization, to put forward Chinese architectural history in order to propagate the traditional Chinese culture, and to manifest the value of Chinese civilization. Liao's architecture matured in northern China during the 10th—12th centuries. It succeeded the style of the Tang Dynasty with its own features—simple yet vigorous and valiant in form with smoother roof gable and higher columns toward the building edges and tilting inward (or, *cejiao*) to support the upward curved oversail eaves and reduce the ponderous feeling usually given by the traditional Chinese curved roof. Especially, its *dougongs* designed under the eaves were simple and solid to serve the load-bearing purpose in contrast with those in the Ming and Qing dynasties mainly for decorations—the older in forms, the more evidence of the rational structural logic of modern architecture are found. In the dusk on June 26, 1937, Liang Sicheng finally successfully discovered the Foguang Temple of the Tang in a deserted rural area of the Wutai Mountain Range. If Liang had found this national treasure before the commencement of the construction project, the Museum would have been designed in reference to it rather than the style of the Liao, for an authentic older Tang style to imitate, the Liao style of a short-lived non-Han ethnic ascendance would not be given a chance.

Another merit in the design art of the Museum is the efficiency of the overall plan and providence.

For such a colossal public building at the national level, it was quite a challenge to design a solemn, grandeur and politically commemorative architecture on a "chopper-shaped" site. From a sketch drawing (Fig. 2-50) of it, Xu's plan demonstrated his elaborated touch, compared with the rest four proposals that placed the main entrance and the mass at the central axis of the "chopper edge" causing the view, from the entrance at the "chopper handle" entering from Zhongshan Lu skewed eastward by 50 meters, blocking the view of appreciating the solemn and grandeur effect that was intentionally created with the symmetrical mass of the overall plan of the designer and making any axial manipulations ineffective to reduce and redirect the first impression of perplexity caused by the "chopper" shape. Tong's proposal revealed this more obviously—"failing to deliver any solemnity at all at the main entrance". Whereas Xu's design used different approach by placing the main building and the entrance on the central axis of the "chopper handle" and the three halls in the north and the west that emphasized the alignment of the main building, the entrance and the internal association of phased construction and land blocks. Or, should we say it was the providence of the "Construction Commission" and the accentuation contrasted by the rigid symmetry of other proposals, which distinguished the vantage and foresight of Xu's "organic symmetry" design that has been proven by the renovation and expansion of the Museum lasting for more than 60 years, although the author doesn't agree with the recent practice of jacking the entire hall.

Case 2 Shanghai Special Municipal Government Building

Shanghai Special Municipal Government Building, situated at 650, Qing Yuan Huan Lu, Yangpu District, Shanghai, occupying a construction area of 9,000 square meters, is now the Administration Building of Shanghai University of Sport (Fig. 2-51).

Shanghai City Central Region Construction Committee was organized by the National Government in July 1929 to prepare *The Greater Shanghai Plan* and *The Great Shanghai City Center Administrative Region Map*. The City Hall was planned to be at the central area of Wujiaochang. In October of the same year, the Committee started soliciting openly the design drawings and set the expert review panel to include Ye Gongchuo, Henry K. Murphy, Dong Dayou and Hans Berents. The design submitted by the well-know architect Zhao Shen, won the first prize. However, for reasons, it was not adopted. The consulting architect Dong Dayou of the Committee redesigned the master drawing according to the first three proposals to let the designers Zhao Shen, Wu Zhenying and Fei Libo to join the review panel to select the final plan. Dong was appointed to preside over the technical design and construction project. In 1931, The Zhu Sen Kee Construction won the construction tender. The construction commenced in June of the same year. Mayor Zhang Qun held the foundation-laying ceremony. The project was interrupted by the "January 28 Incident" and was finally completed on October 10, 1933. The total cost was 750,000 yuan (fiat currency of Nanjing National Government issued in 1935). The building inauguration ceremony was on the 22nd anniversary of Xinhai Revolution (or the Hsin-hai Revolution, also known as the Revolution of 1911 or the Chinese Revolution). The whole Shanghai City was off to celebrate. Guests of home and abroad and over 100,000 spectators joined the ceremony in front of

Fig. 2-51 Shanghai Special Municipal Government Building

Fig. 2-52 The Inauguration Ceremony of Shanghai Special Municipal Government Building

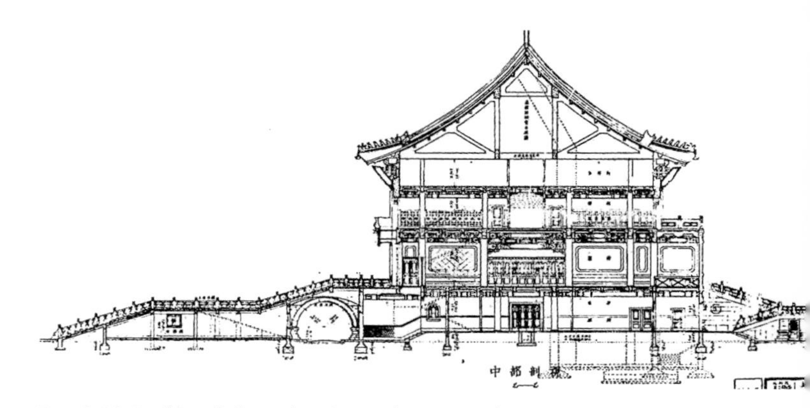

Fig. 2-53 Profile of Shanghai Special Municipal Government Building

Fig. 2-54 Detail I of Shanghai Special Municipal Government Building: door hood

Fig. 2-55 Detail II of Shanghai Special Municipal Government Building: pediment of the hip-and-gable roof

the square of the City Hall, when nine-fighter aircraft squadron did a flypast (Fig. 2-52). The incumbent head of the Department of Public Works of the municipal government awarded Dong a copper medal of "Inauguration of the Shanghai Municipal Government Building" with an inscription, "To Architect Dayou, Ingenuity Award, by Shen Yi", on the back. On January 1, 1934, new Mayor Wu Tiecheng and other municipal government officials moved into the building. On April 3, 1935, the first collective wedding ceremony of Shanghai City, a grand event of Shanghai's civil activity, was held here. After the outburst of "813 - Battle of Songhu", the southeast corner of the building was damaged by bombardments. After the fall of Shanghai, the City Hall was occupied by the Imperial Japanese Army and later used as the office of the Reorganized National Government of China (a puppet government led by Wang Jingwei, also known as the Wang Jingwei regime). After the China's victory of Sino-Japanese War (1937–1945), the National Government moved into the original concession region. The original City Hall was changed to National Sports College, later changed to East China Institute of Physical Education after 1949, and settled as Shanghai

Fig. 2-56 Detail III of Shanghai Special Municipal Government Building: flagpole base

Institute of Physical Education in 1956.

Ferroconcrete frame structure was applied to the main body of Shanghai Special Municipal Government building while ferroconcrete truss structure was applied to its "traditional Chinese curved roof" (Fig. 2-53). The building is four-story high. The main entrance is at the first floor. One door each is in the front, rear, east and west sides. There are two large stairways and two elevators to the fourth floor. The first level has a gatehouse, safe vault, reception room, dining hall and kitchen. The second level has an assembly hall, library and conference room. The mayor's office and offices of other high rank officials are at the center of the third level, and the two sides are offices of sections. The fourth level has the janitor's room, storage room, archives, and telephone operator room. All rooms in the building were equipped with electrical fans, and hot water pipes for heating purpose, which could maintain a room temperature of 22°C during winter, then. Every level has toilet, lavatory and fire-prevention equipment. It was one of the most advanced governmental buildings in China at that time. A square was paved in front of the building, which was used for parades and city assemblies. In the

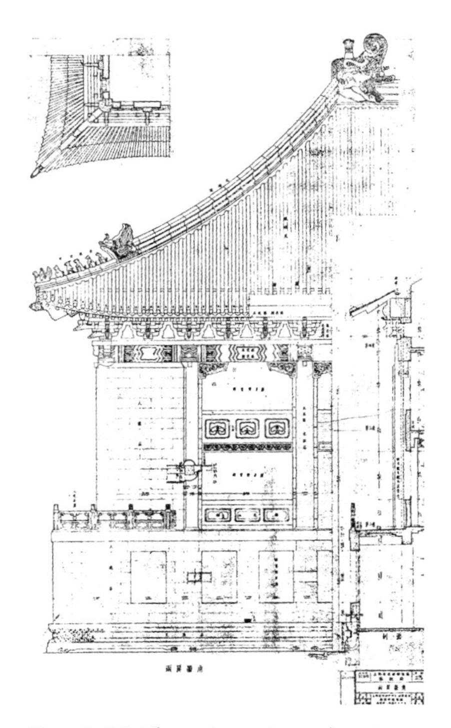

Fig. 2-58 Elevation view showing a concealed gutter in partial elevation plane, wall section and corner eaves of Shanghai Special Municipal Government Building

Fig. 2-57 Interior of the dining hall in Shanghai Special Municipal Government Building

Fig. 2-59 Downspout mouth of a concealed gutter of Shanghai Special Municipal Government Building

Fig. 2-60 Gutter under eaves of Guangzhou Sun Yat-sen Memorial Hall (dark openings under eaves)

south of the square, there is a rectangular fountain and a five-gateway memorial arch (or, *pailou*) as the main gate of the City Hall. There is a pool on each of the east and west sides and an entrance gate at the end of each pool, which are used as the east and west gates for the administration area. Two open galleries used for exhibiting arts are at the two sides of the square. In the north of the building is a Sun Yat-sen Memorial Hall, used as a public assembly, where a bronze statue of Dr. Sun Yat-sen is in front of the hall surrounded by trees and flowers to deliver elegant scenery. As the exterior appearance borrows Qing's features of magnificence and solemnity, where crimson columns, wood door hood, *xuanzi* color paintings, *dougongs*, *quetis* and

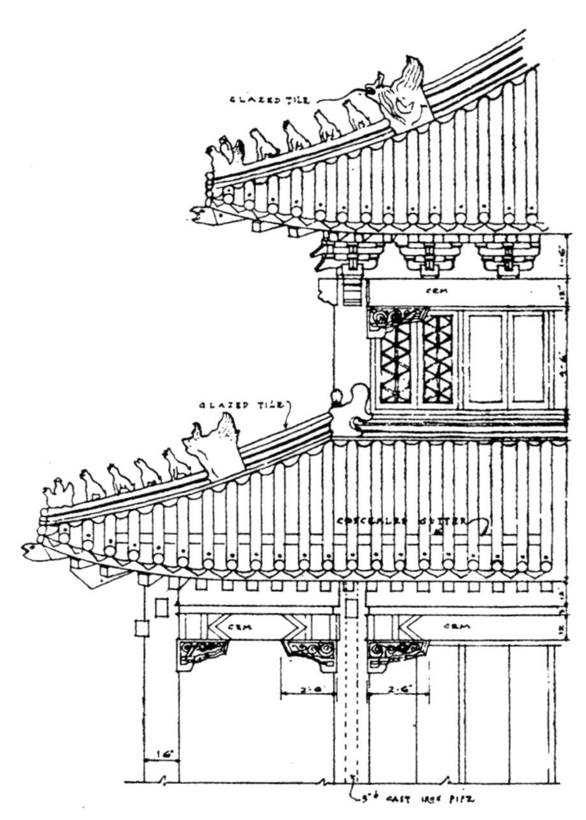

Fig. 2-61 Partial elevation view of the Songfeng Pavilion of the National Revolutionary Army Memorial Cemetery, showing the gutters (legend "GUTTER" above the capital under eaves)

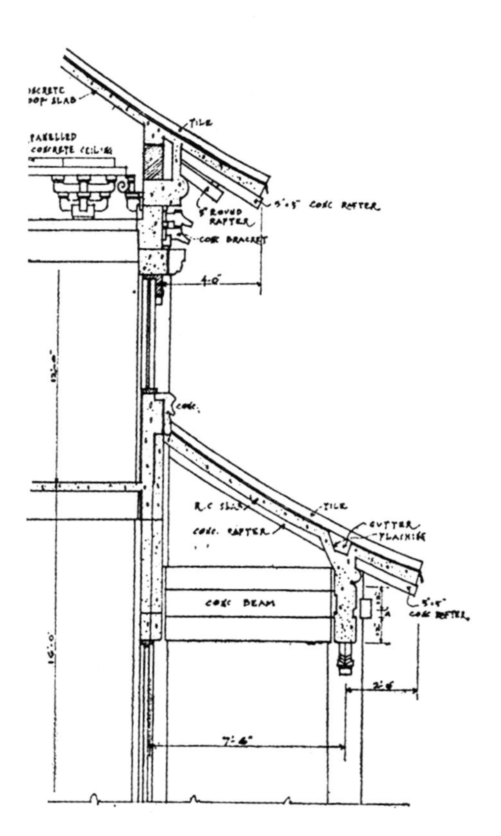

Fig. 2-62 Wall section of the National Revolutionary Army Memorial Cemetery, showing the gutters (legend "GUTTER" above the capital under eaves)

chiwens are implemented in their full scale, including even the flagpole outside (Fig. 2-54 to Fig. 2-56), the interior design follows the usual idea of the integration of Chinese and Western to have a long western dining table integrated with Chinese round-table deisgn, decorated ferroconcrete columns with color painting, and a coffered ceiling (Fig. 2-57), flaunting its extravagance without reservation.

Two new attempts for the art of architecture were undertaken for the building. One—concealed gutters (or, *wogou*)(Fig. 2-58)—the rainwater is collected into the concealed gutters not letting it free fall from eaves onto the ground but sending to the downspouts hidden inside the two sides of the columns (Fig. 2-59) to avoid the water dripping disturbing activities around the main entrance area and also to protect the ground pavement from water erosion. Unfortunately, because of lacking maintenance, the concealed gutters are clogged by dust and weeds. This feature, at that time, was only used on Guangzhou Sun Yat-sen Memorial Hall (Fig. 2-60), the Songfeng Pavilion (Fig. 2-61, Fig. 2-62) of the memorial hall of the National Revolutionary Army Memorial Cemetery, and the "Constabulary Within" of KMT inner court service agency.

Another innovation was the conversion of "inter-*dougong* panel" into sunlight window to effectuate the structure height and space of the traditional Chinese curved roof for natural sunlight and ventilation. The inter-*dougong* panel, similar to *queti,* is a decoration used in between *dougongs* but a connecting panel instead of two independent *quetis.* Designs implemented before were window openings on the pediment of the hip-and-gable roof, such as Ginling College, or a dormer on the front roof, such as Lingnan University. The former achieves limited lighting effect and the latter is not Chinese classical style, while both have deficiencies. Shanghai Special Municipal Government Building has elaborate design on its inter-*dougong* panel to give natural lighting while retain the traditional taste of the hip-and-gable roof. However, not to its perfection, these glass windows on inter-*dougong* panels cannot open to satisfy the natural ventilation needs (Fig. 2-63).

Fig. 2-63 Glass windows on "inter-dougong panels" under the eaves of Shanghai Special Municipal Government Building

Fig. 2-64 Main entrance to the National Revolutionary Army Memorial Cemetery

Fig. 2-65 The Memorial Arch of Benevolence and Love in the National Revolutionary Army Memorial Cemetery

Case 3 Nanjing National Revolutionary Army Memorial Cemetery

Nanjing National Revolutionary Army Memorial Cemetery is situated inside the scenic area of the Linggu Temple encircled within the perimeter of Nanjing Sun Yat-sen's Mausoleum. It was inscribed as a historical monument and cultural relic under state protection (Fig. 2-64).

After the founding of Nanjing National Government in April 1927, to inter and honor military casualties during the National Revolution (1924—1927), a memorial cemetery was under planning. In November 1928, the Central Executive Committee of the Chinese Nationalist Party Kuomintang proposed the "National Revolutionary Army Memorial Cemetery Construction Preparatory Committee" and assigned Chiang Kai-shek, Chen Guofu, Liu Jiwen, He Yingqin, Lin Huanting, Xiong Bin, Liu Puchen, Li Zongren, Qiu Boheng to serve the course. Chiang et al. surveyed the site several times and decided to use the old site of the Linggu Temple to form a tripartite with Sun Yat-sen's Mausoleum and the Xiaoling Mausoleum of Emperor Taizu of Ming, as the core site of Dr. Sun Yat-sen Memorial Park. Henry K. Murphy, an American architect, was hired to design the cemetery, Liu Mengxi as the supervising engineer of the construction and Liang Dingming as the art specialist. The construction project of the cemetery complex was rewarded to Shanghai Voh Kee Construction. The construction was finished in 1936. The total cost was 920,000 yuan.

The plan design aligns the main entrance, the *memorial arch*, the sacrificial hall, the cemetery, the memorial hall, and the memorial tower in order from the south to the north along the south-north axis. The arrangement is according to the layout of the traditional Chinese tomb yet is combined with the western geometrical green square in a subtle integration, contrasting and harmonizing, which forms a scene with spatial depth and manifold inducement.

Main Entrance: Converted from the original gate of the Linggu Temple, the entrance is the enlarged version of it, roofed with green glazed roof tiles supported by three arches and connected by two guard quarters on both sides. On the overhead architrave, there was an inscription rubbing of Chiang Kai-shek hand-written tablet, "Nanjing National Revolutionary Army Memorial Cemetery".

Memorial Arch: An access passage is paved with stone slates leading from the main entrance inward to a large ferroconcrete terrace decorated with granite. At the center of the terrace, a memorial arch with six columns or five gates with 11 towers or 10m high, was built with ferroconcrete imitating Qing's stone-style memorial arch and roofed with green glazed roof tiles (Fig. 2-65). On the top of the entablatures of the five gates of the memorial arch, there is a Kuomintang emblem in vitreous ceramic at the center of each. On the frieze of the main entablature of the central gate, there are light engravings around the edge and an inscription of "Great Benevolence and Great Righteousness" on the front side and another inscription "Heroes of the Nation and the People" on the back side, inscribed by Zhang Jingjiang. The inscriptions are accompanied on each side by a five-pedal blossoming plum design—the national flower of the period of the republic. On the main architrave and other architraves of the gates, there are simplified colored patterns in Qing's style. There is a pair of stone tigers presented by the 17th Army sitting at each side of the main gate of the memorial arch.

Sacrificial Hall: It is the renovated Beamless Hall (or *Wuliang Hall*). The beamless hall was called Hall of Amitayus, where Amitabha Buddha was worshipped. With double-layer hip-and-gable roof covered by clay roof tiles, emitting solemnity and grandeur, the hall has a firm structure built completely with bricks during the Ming Dynasty. When constructing the Cemetery in 1930s, it was restored to its original form with only the inside hall modified as a sacrificial hall. Inside the sacrificial hall, there are three ritual arches with three stone tablets. The central stone tablet has an inscription of "National Revolutionary Army Spirit Tablet" inscribed by Zhang Jingjiang. The tablet on

Fig. 2-66 National Revolutionary Army Memorial Cemetery and Monument

Fig. 2-67 Songfeng Pavilion in the Memorial Hall of the National Revolutionary Army Memorial Cemetery

Fig. 2-68 The National Revolutionary Army Memorial Tower

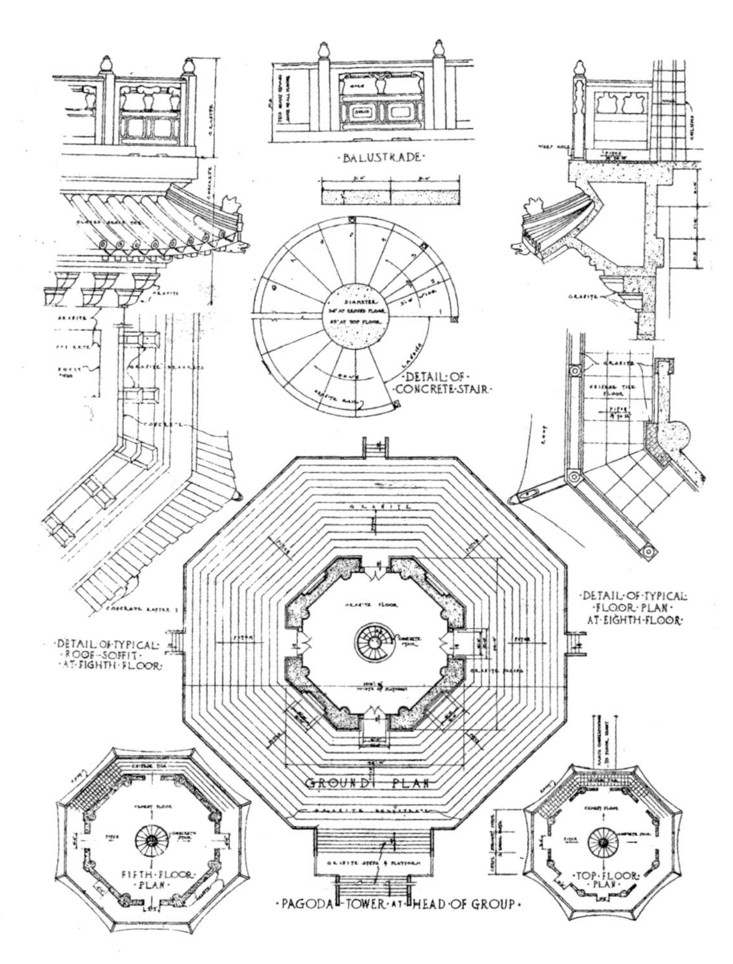

Fig. 2-69 Ground plan and details of the National Revolutionary Army Memorial Tower

the west has the rubbing of "Northern Expedition Pledge" from Chiang Kai-shek. The tablet on the east is the inscription of the funeral oration from Chen Guofu. Four walls on the sacrificial hall are mounted with 110 stone slates, which have over 30,000 of names and ranks of the sacrificed officers and soldiers during the revolution. The entire hall delivers solemnity. It is an exemplar of renovation of the traditional Chinese architecture for modern use during the Republic of China era.

Cemetery: The first cemetery, in the north of the sacrificial hall, situated on the relics of Wufang Hall of the Linggu Temple (Fig. 2-66), has 1,600 catacombs arranged in large, middle and small sizes accessible with webbed passages. Two cenotaphs erect on each side of the east and the west side of the first cemetery for soldiers killed in Songhu Anti-Japanese War (January to April 1932) of the 19th Army and the 5th Army of the National Revolutionary Army. The ground of the first cemetery is now renovated and covered by floral beds and a lawn patch. The second and the third cemeteries similar to the first cemetery in form and area are in a corrie of 1,000m in length on the east and west sides of the Beamless Hall. The three cemeteries were built by Nanjing Li Xin Kee Construction for a

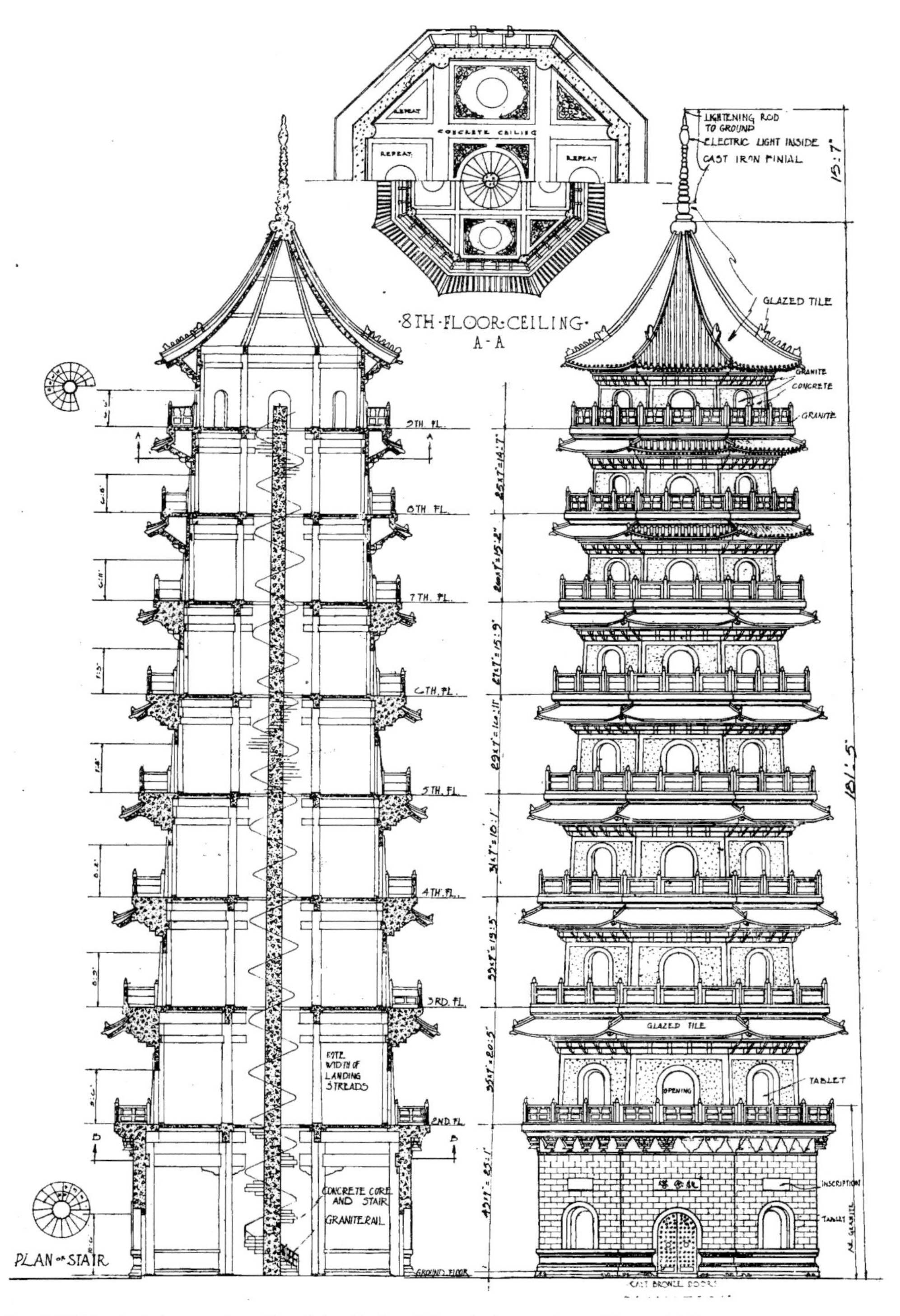

Fig. 2-70 Vertical view and profile of the National Revolutionary Army Memorial Tower

Fig. 2-71 Interior of the first floor in the National Revolutionary Army Memorial Tower

cost of 170,000 yuan.

Memorial Hall: Situated in the straight north of the first cemetery, in nine-bay double-layered hip-roof top covered with green glazed roof tiles and ferroconcrete-imitated wood structure, the memorial hall has two stories. It cost 215,000 yuan then. The two-story space is full of exhibition cabinets used to display survival items of passed soldiers or for other exhibitions. The hall was renamed Songfeng Pavilion (Fig. 2-67) in 1949.

Memorial Tower: About 100m in the north of the Memorial Hall, there is an octagonal nine-story tower-type pagoda designed by Murphy and Chinese architect Dong Dayou, and built by Shanghai Voh Kee Construction (Fig. 2-68 to Fig. 70). The octagonal terrace at the tower base is fenced by a set of carved balusters, and it can be accessed by four-side stairways. At the center of the front stairway, there is a white granite large slab with a meticulous relief "National Landscape under the Rising Sun". There are eight doors—four-hidden and four-vivid—arranged with generous spacing on each story of the tower. The tower has a tapered body of 60 meters in height. Each story is surrounded by a gallery and a set of balusters allowing scenic overlook. A ferroconcrete spiral stairway is placed in the center of the tower. Each story has skirting eaves covered with green glazed roof tiles. Except the top story, each story has stoned tablets with inscriptions of funeral orations of dignitaries of the National Government inlaid on the interior and exterior walls. The tower was renamed Linggu Pagoda in 1949. It is the landmark of the Linggu Temple scenic spot and the Zhongshan scenic area.

Assessed from the architectural design and artistic effect, the tower has a combination of innovation and blemish. The color painting in the entrance hall is a masterpiece of the "contemporarily styled Chinese classical architecture" of the republic era. The mastery is in the simplicity and pertinency of the traditional "originality"—retaining only the geometric "hoop" and having the *Xuanzi* motif and other thematic patterns removed, which delivers direct and simple visual pleasance in modern abstract painting relish yet not curt and pert (Fig. 2-71). Compared with the work done 10 years ago for Ginling College, Murphy had made significant progress in grasping the "originality" of the art of Chinese palace architecture. However, the

Fig. 2-72 Perspective view of the main administration building in the Academia Sinica, Nanjing

Fig. 2-73 The main administration building in Nanjing National Academia Sinica with typical characteristics of the "Contemporarily Styled Chinese Classical Architecture"

pagoda has an evident blemish in that the design of "entasis" of the pagoda's inward inclination is a linear form not displaying an elaborate touch of "entasis". The shape of the overall pagoda appears rather bored stiff, lacking the dynamics of a traditional Chinese pagoda.

Case 4 The Academia Sinica in Nanjing

The original Academia Sinica is at 39, Beijing Donglu, Nanjing (the old address 1, Jiming Temple Road). The complex includes the main administration building, Institute of Geology, Institute of Social Science Research, Institute of History and Philology. In 2005, it became a historical monument and cultural relic under state protection (Fig. 2-72).

The Academia Sinica (June 1928—April 1949), directly under the state sponsorship, was the highest academic research institution of China. Its major mission was to conduct scientific research, direct, coordinate and commend scientific achievements. It consisted of administration, research and evaluation institutes. Its research institutes were organized by categories. By the outburst of Anti-Japanese War, there were ten institutes of physics, chemistry, engineering, geology, astronomy, meteorology, history and philology, psychology, social sciences, botany and zoology. Except physics, chemistry and engineering institutes were located in Shanghai, the rest were all in Nanjing. The academic senate of Academia Sinica was the highest academic senate during the republic era. In

Fig. 2-74 Porticos used as main entrance and exit in the main administration building of the Academia Sinica, Nanjing

April 1948, 81 academicians were elected and staffed. In the same year, the Academia Sinica was moved to Taiwan with the government of the Republic of China, when nearly 20 academicians followed the retreat or went overseas. In October 1949, the rest academicians remained in China were transferred to the Chinese Academy of Sciences (CAS).

The Administration Building of Academia Sinica is now used by Jiangsu Science and Technology Department and CAS Jiangsu Branch. The building facing south was built in imitated Qing's official building style. The two sides of the main gate and the east side of the encircling wall have three guard houses with single-layer tetragonal prism roof, similar to the main building. The building was designed by Yang Tingbao of Jitai Construction and constructed by Xinjinji Kanghao Construction in 1947. It has three stories and a construction area of 3,000 square meters. Its main body is in ferroconcrete structure with single hip-and-gable roof covered by green glazed roof tile. Beams, architraves and cornices are all in imitated wood structure with color painting. The exterior wall was built with Taishan bricks. With tracery doors and windows, the building appearance displays the typical characteristics of the "Contemporarily Styled Chinese Classical Architecture" (Fig. 2-73). The building plan is an inversed "T" shape. The entrance has a two-

Fig. 2-75 Exquisitely executed connection I between wall and roof in the main administration building of the Academia Sinica, Nanjing

Fig. 2-76 Exquisitely executed connection II between wall and roof in the main administration building of the Academia Sinica, Nanjing

Fig. 2-77 Roof details of the main administration building in the Academia Sinica, Nanjing: abstract and simplified deity and beast statuettes themed with cloud patterns

Fig. 2-78 Space utilization of the hip-and-gable roof in the main administration building of the Academia Sinica, Nanjing: window opening on the pediment

story vestibule with a decorated door. After passing the vestibule to the leg of "T", there is a three-storied library. The west wing of the front building has a small assembly hall used for academic activities. The front building has a hallway where its north and south sides are used by offices and research facilities.

The rest space of the Administration Building is used by institutes of Geology, History and Philology and Social Sciences, which were designed by Yang Tingbao and built in 1930 before the Anti-Japanese War. It is 2—3 stories high, with the exterior appearance in "China's inherent architectural style", structured in ferroconcrete topped with a section of brick walls without applying plastering and bottomed with artificial stone plinths covered in granitic plaster.

As the most significant building of the Academic Sinica complex, the physical construction of the Administration Building is larger than those of the other three and more magnificent in appearance. Its features are:

1. Although the plan is a simple inversed "T" shape, the application of the "traditional Chinese curved roof" of its "contemporarily styled Chinese classical architecture" is surprisingly rich. The central ridge has a hip-and-gable roof on its second story topped by another overhanging gable roof on its third story. The two sides are lower and capped by flat roofs. The entrance at the central south is a "portico" topped with a hip-and-gable roof orthogonally connected with the building. The layered sensation is vivacious. The entrance opening aligned with the façade is the typical method used in the western classical architecture while being innovatively integrated with the "portico" of "Chinese palace architecture" (Fig. 2-74).

2. The façade is a brick wall without columns. Every room has two window openings articulated with half-brick stacks in between to unveil the inner room size. Between stories on the exterior wall a horizontal "frieze" in artificial stone is in continuity at the same elevation with the parapets on the top of the two-side flat roofs. The design delivers appearance simplicity of a modern office building. The architect's ingenuity is shown in the connection of the wall and the roof, outwitting the difficulty of integrating the modern external wall and the "traditional Chinese classical curved roof"—especially, the details under eaves, which evades deadening and patching gimmicks. The top of the external wall does not use open cornice but similar wide-box cornice with color painted molding. By adding single brackets, eave purlins and chapiters, the design expresses a rational structure of the combination of chapiter, *dougong* brackets, beam and architrave, with which it forms a natural harmonious

Fig. 2-79 Perspective view of the National Wuhan University

transition avoiding the stiffness of a close cornice (Fig. 2-75, Fig. 2-76). Comparatively, the handling of these details in earlier "contemporarily styled Chinese classical style" was immature and roughly done.

3. The brilliant part of the detailed decorations is in the simple presentation of the "originality" of the traditional elements and the well balanced abstraction, such as deity and beast statuettes on the ridges are all reduced to abstracted cloud patterns not in the actual human and beast forms. However, when viewing the overall building, the essence of Chinese palace architecture is unveiled completely (Fig. 2-77).

4. On the pediment beneath the bargeboards and above the entablature, the window opening provides natural lighting to the upper part of the structure and increases the rate of space utilization. The design is also very creative. (Fig. 2-78).

In summary, as the closing stage of the "contemporarily styled Chinese classical style" before 1949, the Administration Building of Academia Sinica testifies the modernism transition of the "architectural integration of Chinese and Western" from its ameliorating architectural art and design of similar works earlier.

2.4 Reassessing College Campus Architecture

Since the republic era, higher education has become popular and new campus architecture has been keeping emerging. Before 1920s, it was the active period for the church school construction so was the first wave of "contemporarily styled Chinese classical architecture". The success of the Northern Expedition and the founding of Nanjing National Government gave the impetus to "nationalized" church schools. In the end of 1920s, a new wave of national university campus construction emerged, although a short period, but its impact cannot be ignored. Politically, the architectural activity could be directed by the Chinese alone, especially, after the first generation of Chinese modern architects had emerged into the architecture stage and proved their competence, the reality did not entirely agree with them, however. The planning of Wuhan University was supervised by Li Siguang but the design was done by US architect F. H. Kales (1899—1979). Whereas, Sun Yat-sen University and Xiamen University were planned, designed and directed by Tan Kah Kee (Chen Jiageng). The project on Xiamen University almost was not assisted by any architect but only engineers and craftsmen. Sun Yat-sen University was completed entirely by Chinese architect. Influenced by the demand of "China's

Fig. 2-80 Main entrance to the School of Engineering, National Wuhan University

Fig. 2-81 Rear part of the School of Engineering, National Wuhan University

Fig. 2-82 A corner of the student dormitories in the National Wuhan University

Fig. 2-83 Detail of the entrance to the student dormitories in the National Wuhan University

Fig. 2-84 Ingenious design in harmony with landscape in the profile designing of the student dormitories in the National Wuhan University

inherent architectural style" of that era, the three universities have different characteristics that deserve assessment from other perspectives.

Case 1 National Wuhan University

The original National Wuhan University sitting at the foothill of the Luojia Mountain Range and the lakeside of East Lake of Wuhan City is now Loujia Campus of Wuhan University. The old site is richly endowed by natural scenery—abutting a mountain and on a lakeside. The building, different from then popular large public punctilious "palatial" style, is an exemplar of the contemporary Chinese campus

Fig. 2-85 Exterior of the National Wuhan University Gymnasium

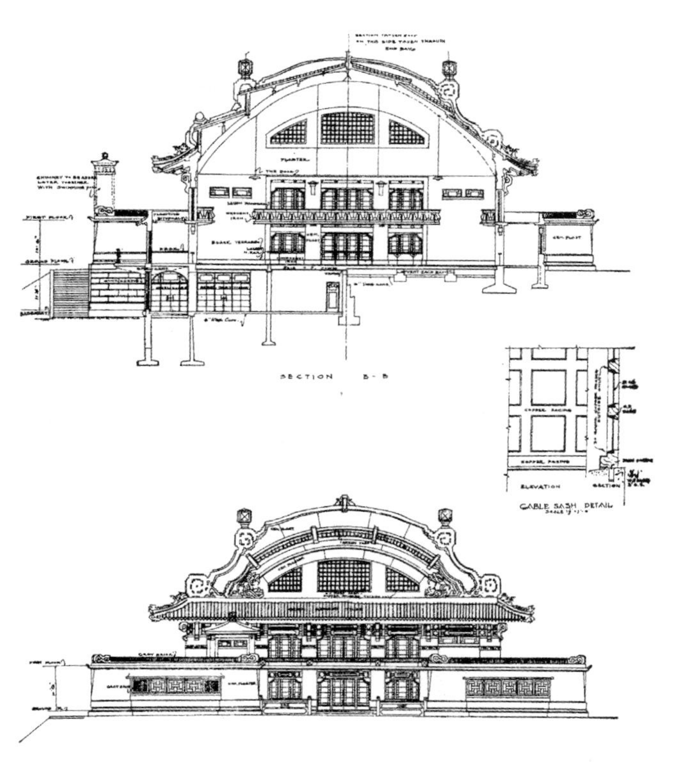

Fig. 2-86 Profile and side elevation of the National Wuhan University Gymnasium

architecture. Its building complex is exquisite in layout, unique and innovative in design, emitting "China's inherent architectural style" from its western style building body. In June 2001, the complex was designated as one of the fifth group of historical monuments and cultural relics under state protection (Fig. 2-79).

In July 1928, under the endorsement of Cai Yuanpei, director of the department of Nanjing National Government Colleges and Universities (the Ministry of Education), the original Wuchang Sun Yat-sen University changed its name to National Wuhan University. Cai proposed the geologist Li Siguang to chair the "National Wuhan University New Campus Building Preparatory Committee" to preside over the construction. Li and Ye Yage were to select the

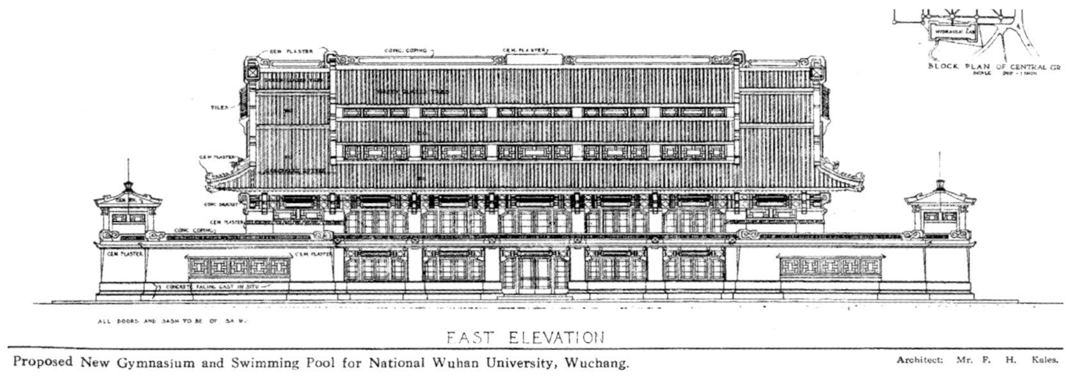

Fig. 2-87 Front elevation of the National Wuhan University Gymnasium

Fig. 2-88 Interior of the three-hinged steel arch structure in the National Wuhan University Gymnasium

Fig. 2-90 Details of the roof of the National Wuhan University Library

Fig. 2-89 The National Wuhan University Library

Fig. 2-91 A corner of the School of Science in the National Wuhan University

campus site, plan and raise funds for the construction project. They hired US architect F. H. Kales to preside over the planning and design and Miao Enzhao to be responsible for the supervision of technical and structure work. Han Xiesheng Construction, Yuan Ruitai Construction, Shanghai Liuhe Construction and Yong Maolong construction split the construction contract. The project was commenced in March 1930 and completed in 1936. The major buildings included in this project were schools of Liberal Arts, Law, Science, Engineering and Agriculture, and amenities, such as library, gymnasium, student dormitories, teacher dormitories, student cafeteria, clubs, laboratories, workshops, main gate (*paifang*) and water

Fig. 2-92 Details of the School of Science in the National Wuhan University

towers. The campus occupied an area of 3,200 *mu* and the construction area occupied over 70,000 square meters. The total cost was more than 4 million yuan.

The beauty of the campus architecture of Wuhan University gaining its reputation at home and abroad is due to:

1. Campus Site: The campus site selected by Li and Ye is not only a natural but also culturally significant landscape in reference to international renowned college campuses and the geomancy for ancient Chinese schools rooted from the philosophy of "the wise enjoy the waters and the benevolent embrace the mountains". To situate a national high education institute in a beautiful landscape site is a wise choice that integrates the natural scenery and cultural heritage.

2. Campus Planning: Kales lived up to the elite reputation of Massachusetts Institute of Technology (MIT). He integrated Chinese and western architectural merits in coherence with the landscape of the Luojia Mountain Range and modern high education needs to create a brilliant campus layout—dividing campus buildings by functions and arranging them on radiant axes in ordered elevations according to the terrain (Fig. 2-80, Fig. 2-81 and Fig. 2-83). He put the central garden and athletic field at the basin surrounded by Mount Shizi, Mount Huoshi and Mount Xiaogui on three sides and an open valley on the west, and main buildings and other amenities on these three mountains by the topography of mountain shapes[1]—forming a "symmetrical layout along axes in primary and subordinate relationship, where the hall is at the center and storied buildings at corners". These buildings are positioned on dynamic elevations opposing and overlooking each other that enriching complementarily the spatial layers of the natural landscape.

Fig. 2-94 The School of Liberal Arts in the National Sun Yat-sen University (from Zhao Yunfei, South China University of Technology)

Fig. 2-93 Current west gate of the National Sun Yat-sen University (from Zhao Yunfei, South China University of Technology)

Fig. 2-95 The School of Law of the National Sun Yat-sen University (from Zhao Yunfei, South China University of Technology)

1 Li Chuanyi (1982), Preliminary Planning and Construction of Wuhan University, New Architecture (Issue 3, pp 40)

Fig. 2-96 Agronomy Hall, the School of Agriculture, the National Sun Yat-sen University (from Zhao Yunfei, South China University of Technology)

Fig. 2-97 A classroom for biology, geology and geographic students in the School of Science, the National Sun Yat-sen University (from Zhao Yunfei, South China University of Technology)

3. Architectural Design: Each architectural entity plays its role according to the elevation of the terrain, for example, the student dormitories (Fig. 2-82, Fig. 2-84); space manipulation by innovation ideas, for example, the courtyard of the school of engineering uses glass roof to form a shared space; new techniques adopted aggressively to deliver spatial functions, for example, the gymnasium uses three-hinged steel arch structure (Fig. 2-85 to Fig. 2-88); and, the library combines the ferroconcrete frame and modular steel truss structure. Although, these new space concepts, structures, materials and technologies were still under experiment, some were mature but newfangled in China; however, they were successfully implemented in the campus construction. Besides this, in style and appearance, the design of Wuhan University had integrated the traditional Chinese curved roof and the classical western dome style. The elevation façade has "three sections". The main body is expressed with impressions of Chinese official city gate and stylobate. The exterior wall follows the proportion and scale used in classical western stone design with window openings. The interior details preserve the originality of the traditional Chinese wood structure in unique abstractions or modifications (Fig. 2-89 to Fig. 2-92). This approach makes this campus architecture not only to have characteristics of Chinese wood structure in style and format but also to present the verve of the western stone architecture, delivering a new look of the traditional Chinese architecture and bringing forward the architectural integration of Chinese and Western. It stands out among the China's contemporary college campuses.

Case 2 National Sun Yat-sen University

The campus of National Sun Yat-sen University, founded in 1930, is now South China University of Technology (SCUT) and the Shipai Campus of South China Agricultural University (SCAU). In 2002, the site was designated as one of Guangzhou Priority Protected Sites (Fig. 2-93). In early 1924, Dr. Sun Yat-sen, as Commander in Chief of the Army and the Navy, ordered the construction of the first multidisciplinary college by Chinese—National Sun Yat-sen University that later was expanded as a large comprehensive university in South China. During 1932 and 1937, to fulfill the will of Dr. Sun Yat-sen, Shipai Campus of National Sun Yat-sen University was built. When the outbreak of the Anti-Japanese War, the new campus was finished and praised for its scale, facilities and architecture.

During 1927 and 1930, Dai Jitao, as president of the university, started the founding plan of the university, but failed to substantiate it. Later, Dai and the renowned architect Yang Xizong jointly did the campus layout[1] according to criteria of "Bi Yong" (Imperial School of the Western Zhou Dynasty, which had circular architecture resembling heavenly fullness). Dai considered that the campus architecture should be solid and practical and decided to build the auditorium, library and museum first. The funding hindered the campus construction until in 1932 when Zou Lu succeeded the presidency of the university again. Riding on the blooming economy of Guangzhou, Zou launched the construction project with unanimous supports from the incumbent chairman of Guangzhou Chen Jitang and all sectors of the society. He made a few modifications from the previous design and divided the project into three phases. Yang Xizong was responsible for the planning and design of the first phase, Lin Keming for the second and Yu Qingjiang for the third. Zou presided over all three phases. In six years, before 1938 Guangzhou Siege by the Imperial Japanese Army, the construction plan was completed. Unfortunately, the successive warring chaos stopped the initial construction of the library before key buildings of auditorium and museum could even made to the agenda. The Shipai Campus was built over 70 years ago. The site selection, planning and design of the overall architecture reveal some evident features:

1. Prudence in Site Selectin and Cultural Revival by Landscape Integration

Occupying an area of 180 hectares, adjacent to the Pearl River Delta in the south and in 10 km distance from the city center of Guangzhou, the site of Shipai Campus was selected in the east of a hilly land patch of Guangzhou Baiyun Mountain Range. It was originally a barren mountainous terrain and later used as the second cropland by the School of Agriculture. In 1928, Guangzhou Municipal Government allocated additional abutting 400 hectares to the campus. In

Fig. 2-98 The National Sun Yat-sen University Gymnasium (from Zhao Yunfei, South China University of Technology)

Fig. 2-99 Awning at entrance resembling western classical stone columns in the School of Liberal Arts, the National Sun Yat-sen University (from Zhao Yunfei, South China University of Technology)

1936, the school expanded to 806.7 hectares.[2] Using this land as the school campus avoided farmland acquisition and lowered the cost, while also giving the school sufficient space to grow. Compared with the rush of new campus construction in remote suburban

1 Zheng Lipeng (2004), *Paradigm of Campus Consturction of National Universitise in Contemporary China—Campus Planning and Construction of Old National Sun Yat-sen University*, New Architecture (Issue 6, pp 64—67)

2 ditto

Fig. 2-100 Awning at entrance with a flat parapet covered with green glazed roof tiles in the School of Liberal Arts, the National Sun Yat-sen University (from Zhao Yunfei, South China University of Technology)

areas flocked by high educational institutes of which some are rather in debt under the tension imposed by land availability in city center, "enrollment expansion" and "merger" in recent years, the strategic deliberation of the site selection of Shipai Campus has proven its planners' prudence.

The Shipai Campus site was an undulating land. The plan layout must stay in harmony with the landscape outline to avoid unnecessary construction efforts. During the implementation, not only the bell-shaped road map planned in the layout of 1930 was changed but also the building plots to adapt to the terrain—using buildings and roads to decorate the hills and form three north-south topographical axes, and formulating a east-west water system by filling valleys and by building banks for pools and ponds in different sizes. Therefore, the original hilly barren was transformed into a landscape campus.

As the first comprehensive college in South China founded by the Chinese, the National Sun Yat-sen University was founded in early 1932 right after the "Mukden Incident" (also known as the Manchurian Incident) when the public patriotism was at its peak. The campus planning emphasized, especially, on nationalism and patriotism to build a "Chinese College Campus" epitomizing the traditional Chinese cultural ideology. The campus sitting on the north-south axis with the School of Agriculture positioned at the north-end to represent the Chinese "agriculture-oriented" ethos. The rest school buildings were symmetrically situated along the axis, where the School of Liberal Arts and the School of Science were on the east side and the School of Law and the School of Engineering on the west to manifest the traditional doctrine of ritual precedence—"Intellectuality on Left and Military Might on Right". The auditorium was placed at the center, flanked by the library and the museum. The experimental cropland and the forest were situated on the left and right sides of the campus central region. All buildings were designed in "China's inherent

Fig. 2-101 Bird's eye view of Xiamen University from a peak

Fig. 2-102 A view of Xiamen University from the sea

architectural style" (Fig. 2-94 to Fig. 2-98). Within the campus perimeter, all land parcels, hills, ponds, pools, and roads were positioned and named after Chinese provinces, mountains and lakes to motivate the patriotic enthusiasm of teachers and students.

2. Integration of Nationalistic Renaissance and Western Technology

The campus was planned and designed by the returned Chinese architects trained abroad. Yang Xizong (1889—?, studying along with Lü Yanzhi in Cornell University) won the third place of Nanjing Sun Yat-sen Mausoleum design competition, and Lin Keming (1900—1999), a Chinese architect studying in France, was the construction consultant of Guangzhou Sun Yat-sen Memorial Hall and the designer and the tender-winner of Guangzhou Library and Guangzhou Municipal Government Administration Building. While, Yu Qingjiang (1893—1980), a self-study apprentice of a design house, turned architect without a diploma in architecture, dedicated his life to the profession with outstanding achievements. The three designers from different educational backgrounds shared the same design idea. They, when designing the site, integrated the western modern architectural ideas, technologies and materials with the traditional Chinese cultural renaissance ideology to satisfy the requirements of civilization advancement and functional performance. The important buildings were designed in "China's inherent architectural style" to propagate the traditional Chinese culture. The simplicity of western modern architectural style was adopted for the rest buildings. The buildings were arranged by functional weight. The Chinese and Western styles co-existed with concordance on campus. Every building was designed with functional rationality and to be economic and practical. Most of materials used were then new materials and technologies, such as ferroconcrete frame structure, steel roof truss or ferroconcrete flat roof and steel windows. For example, the School of Liberal Arts

Fig. 2-103 Lotus No.1 Student Dormitory in Xiamen University

Fig. 2-104 Front view of Qunxian Building, Xiamen University

Fig. 2-105 Side view of Qunxian Building, Xiamen University

Fig. 2-106 Jiannan Auditorium in Xiamen University

Fig. 2-107 Jiannan building cluster of Xiamen University

(today's School of Public Administration of South China University of Technology) has straight and simple façade with horizontal cantilevered eaves on the top of its exterior wall, if not for the green glazed roof tiles covering the hip-and-gable roof, it would be hardly not to imagine not a splice of details of a modern building. More dramatic effect is in its design idea of the "architectural integration of Chinese and Western" of three-bay western classical stone columns for the entrance portico to support a flat parapet covered with green glazed roof tiles (Fig. 2-99, Fig. 2-100). Only, compared with other buildings on campus (even including Wuhan University and Xiamen University), the curt patching trace becomes obvious. This also proves the implementation difficulty of the "architectural integration of Chinese and Western style".

Case 3 Xiamen University

The early campus of Xiamen University was at the south end of Xiamen Island. It is today's main campus of the university (Fig. 2-101, Fig. 2-102). In 2006, the site was designated as one of the 6th group of historical monuments and cultural relics under state protection. The founder and the designer was the renowned overseas Chinese Tan Kah Kee.

Tan, born in Jimei of Xiamen in 1874, moved to Singapore to learn business with his father at age 17. After over 20 years hard work, he became a reputed entrepreneur in Southeast Asia. After the success of the Xinhai Revolution 1911, Tan returned home in 1912, with the aspiration of "promoting education in China", and started his journey of devoting to education. In 1919, he planned Xiamen University and raised funds for the school and its perennial reserves. In April 1921, Xiamen University was officially open. In 1936, the university had liberal arts, science and law schools with nine departments in total. During the Anti-Japanese War, the university became the barracks of the Imperial Japanese Army, which caused damage to the campus buildings. After the Japanese was defeated, the school was repaired and renovated but only to be damaged again by the civil war. In the end of 1949, Tan decided to rebuild the Jimei school complex and

Fig. 2-108 Porticos at the entrance to Jiannan Auditorium, Xiamen University

Fig. 2-109 Details of roof combination of Jiannan Auditorium, Xiamen University

Fig. 2-110 Details of a room corner in Qunxian Building of Xiamen University

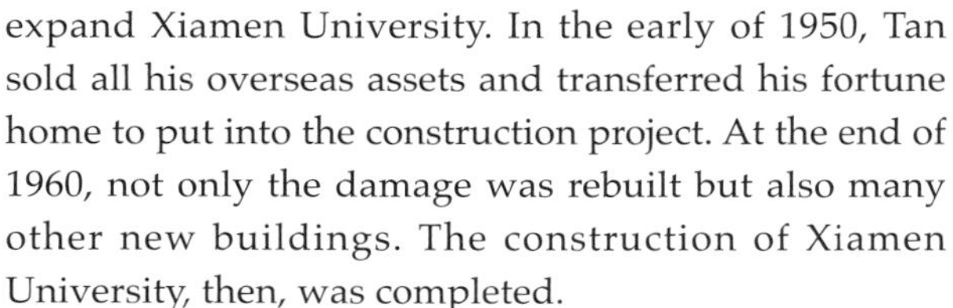

expand Xiamen University. In the early of 1950, Tan sold all his overseas assets and transferred his fortune home to put into the construction project. At the end of 1960, not only the damage was rebuilt but also many other new buildings. The construction of Xiamen University, then, was completed.

The campus architecture evolution of Xiamen University can be divided into three stages. The early stage was between 1913 and 1916. The building drawings were all done in Singapore; the design was the "cloister format" of the colonial style of Southeast Asia and the western architectural classicism (Fig. 2-103). The expansion stage was between 1916 and 1927. With sufficient funding and accrued experience from the early stage, the construction had more flexibility in site selection and layout. The design at this stage adapted the layout to the terrain to emanate the glamour of the building and the building cluster. The integration was shown in its Chinese style roof and the Western building body (Fig. 2-104 , Fig. 2-105). The final stage of "Kah Kee Style" was between 1950 and 1960. The planning and layout added weight on the topography of the terrain. The traditional Chinese curved roof and the western building body became the characteristics of "Kah Kee Style" (Fig. 2-106). In general, the basic philosophy of "Kah Kee Style" is to organically integrate the traditional Chinese architecture with the western architecture, construction technologies and materials. The individual building was done in this way. To the cluster, the building with the traditional Chinese curved roof was placed at the center and buildings with the western roof served as the subordination. The central building was more massive and grandiose.

For example, the Jianan building cluster constructed from 1951 to 1954, including the Chengyi, Nan'an, Nanguang and Chengzhi buildings and Jiannan Auditorium (Fig. 2-107), arranged in crescent shape, was situated at the hilltop of the southeast corner of the coliseum (relics of coliseum built by Zheng Chenggong), of which the Jiannan Auditorium is at the center and the rest four buildings in western style are placed at two sides on a plan layout of three layers. The buildings all have inner hallways, stone and wood structure, granite walls and red tiles in patches on wood floors. The two ridges roofs have red flat-tile shingles. The pediments have arch windows. The corner columns are decorated with mixed "bricks and stones". The front of the Jiannan Auditorium is a portal and the back is the main hall of the auditorium. The building is in stone and wood structure. The portal has four levels. The total construction area is around 5,600 square meters. The roof has double-curved ridges and double-layer-eaves hip-and-gable roof. The ornamental piece at the end of the ridge is in dove-tail shape. The roof is covered with green glazed roof tiles. The pediment and eave downside are decorated with the traditional southern Fujian carved mortar statuettes and wood lotus pendants. The first level of the portal is five-bay in size, supported by four Ionic orders (Fig. 2-108). Right behind the portal is a granite three-bay lobby with its facing façade decorated with fine carved diabase "trims". The lobby has three amaranth double-doors leading to the auditorium hall. The auditorium hall only has one level topped with western-style gable covered with flat roof shingles. The cluster is the masterpiece of "Kah Kee Style", which is also the landmark of Xiamen University.

What is more stunning is that Tan personally involved in the planning, design and supervision of the construction project. He was called the "ultimate chief engineer". Design and supervision of construction work by the investor and owner are not surprising and unprecedented. To create is a human nature. Tan's ideal and enthusiasm were not only unveiled in his investment but also in his "directing" effort of the project—or perhaps, at least qualified to be an "architecture aficionado". The key difference of his style is in the "traditional Chinese curved roof" following closely the southern Fujian style that has curved ornamental piece at the end of ridge so the entire ridge is in an arch form with exquisite decorations, not similar to the preceding "contemporarily styles Chinese classical architectural style" found in North China that are composed and decorous (Fig. 2-109, Fig. 2-110). When these detailed works with traditional folk culture features built by private sectors, the common flat roof shingles, straight ridges, stone walls and the western classical architectural columns are generously combined together. By viewing them, the feeling of crossing time-space gives an amazing sensation. Especially, its imitation does not degrade the academic atmosphere by tawdriness. The successful recipe might be due to the materials and the unpretentious color scheme. As a non-professional architect, Tan's home-soil consciousness, art taste, and design standard were unusual, so was the regional etherealness of Xiamen University that has become the esteemed exemplar of the Chinese modern campus architecture.

2.5 New China: "Traditional Chinese Curved Roof" with "Nationalism Form Integrated with Socialism Content"

The founding of the People's Republic of China and the retreat to Taiwan of the Kuomintang regime began a new epoch of Chinese history. The change of political regime did not affect the culture evolution—not the sprouting process of the architectural integration of Chinese and Western. The persuasion of the theoretical logic of "nationalism form integrated with socialism content" from the former Soviet Union under Stalin's regime to the architectural culture and the significant demonstration effect of the Eclecticism creativeness stemmed from "Beaux Art" to Chinese architectural activities during the special historical "Cold War" era, amplified Chinese nationalism along with the "lean to one side" diplomatic policy of New China. More interestingly, after 1949, across Taiwan Straits, the long-time rivalry civil war standoff did not stop the spreading of Chinese nationalism on either side, or was in non-negotiated coherence. Even to some degree, competitions for orthodox culture and national advancement have been carrying on cross-Straits, and surprisingly, the implementation in architectural activities has stunning resemblance only different in the theoretical ideology and the discourse system.

Fig. 2-111 Exterior of Chongqing Great Hall of the People

Fig. 2-112 Memorial arch at the entrance to Chongqing Great Hall of the People

Fig. 2-113 Main building complex of Chongqing Great Hall of the People in harmony of landscape

Fig. 2-114 Amenities of Chongqing Great Hall of the People in harmony with landscape

While, the "nationalistic style" of the of Literary Theory based on the "nationalism form integrated with socialism content" integrates seamlessly with "China's inherent architectural style"—The "Western" portion of the "architectural integration of Chinese and Western", although coarsely related to "Imperialism" and "Capitalism", which has had its associated political risk resolved under its resplendent cover.

During the epochal changing time, Chinese architecture historians led by Liang Sicheng and Liu Dunzhen and renowned architects, including Yang Tingbao, Zhao Shen and Chen Zhi stayed with New China, which preserved certain talent vantage for mainland China over Taiwan. Amid the reconstruction in the aftermath of warring chaos, the "contemporarily styled Chinese classical architecture" was dominant and reached its prime time at the 10th anniversary of the founding of New China that started from the building of Chongqing Great Hall of the People in 1953 and large constructions before and after the 10th anniversary in Beijing, including the Cultural Palace of Nationalities, Beijing, Beijing Railway Station, National Agriculture Exhibition Center, buildings of four ministries and one national assembly hall, Beijing Friendship Hotel and the Teaching Building of Eastern China Aeronautics Institute (ECAI) in Nanjing, which also served as the milestone of this eventful movement. In the "nationalistic style" aspect, Chinese architects made a series of innovative attempts in integrating the traditional roof style with the ultimately spatial spanning structure, high rises and large-scale building complexes, as well as adapting detailed decorations to the mainstream trend and functional performance of architecture.

Case 1 Chongqing Great Hall of the People

Chongqing Great Hall of the People, situated at Xuetianwan, Renmin Lu of Chongqing City, designed by architect Zhang Jiade, was built between 1951 and 1953. The building was initially named "Sino-Soviet Building" and later changed to "Southwest Administrative Committee Auditorium". In 1955, the name changed again to the name today. The Hall occupies an area of 66,000 square meters of which 25,000 m^2 is the construction area and is divided into auditorium, South Building and North Building. The auditorium has an area of 18,500 m^2 and 65 m in height. The ceiling high is 55 m and the inner diameter is 46.3 m. The auditorium has a large performance stage and five-story audience seating area—four upper levels and 1 base level—accommodating 4,200 seats. In architectural art, the auditorium has three layers of conical roofs resembling the Hall of Prayer for Good Harvest of the Temple of Heaven, a gate tower style entrance and two towers, one in the south and one in the north, supported by red columns. The roofs are covered with green glazed roof tiles. In the front square, there is a three-gate seven-tower memorial

Fig. 2-115 Details of the vigorously-designed main hall in Chongqing Great Hall of the People

Fig. 2-116 Interior of the main hall in Chongqing Great Hall of the People (from Tan Lin, Chongqing University)

Fig. 2-117 Conical top and furred ceiling in double-meshed angle steel trusses used in the main hall of Chongqing Great Hall of the People (From Qin Lin, Chongqing University)

arch. The overall architecture delivers harmony and splendor that can be another exemplar of the "architectural integration of Chinese and Western" (Fig. 2-111).

As the political, economic and cultural center of Southwest China, Chongqing did not have an assembly facility at the early stage of New China. Despite of being shy in funding, the authority still decided to put forward the construction of an auditorium to be able to accommodate several thousands of people and to serve also as a hostel facility. In 1935, architect Zhang Jiade, a graduate of National Central University (founded in 1915, renamed Nanjing University in the Mainland in 1949, re-established in Taiwan in 1962), who was a classmate of renowned architect Zhang Kaiji, presided over the design work. The construction started in 1951 and completed in 1953.

The unique feature of the "architectural integration of Chinese and Western" of the Hall is the western Eclecticism manifested in profile and layout and the expressive vocabulary and details in ancient Chinese wood structure. In architectural technology, it has boldly meshed angle trusses for its conical top. The multitude of optimization of different factors formulates an "ultimate Eclecticism" that is the very characteristic of "Beaux Art".

Restrained by materials, Chinese classical architecture is also limited in volume. The Hall design inherits the past Chinese architecture experience, adapts to the terrain for its building blocks and lays an expansive stylobate to create a magnificent artistic effect (Fig. 2-112 to Fig. 2-114). Furthermore, the Hall uses individual architectural blocks to form a sophisticated complex, i.e., using the building and courtyard formation as a unit to produce a dynamic architectural profile with axiality and spatial order.

The dynamic parallax effect, well dimensioned proportion and beautiful details reveal the sophistication of architectural techniques used by the designer. To proportion the width with the height of the meshed angle truss structure, the main hall, although

Fig. 2-118 Exterior of the Cultural Palace of Nationalities, Beijing

conveying resemblance to the three-layered-eaves conical roof of the Hall of Prayer for Good Harvest of the Temple of Heaven, has moderate sloping stretched out in size but not in elevation. The vertical distance between the three-layered eaves is shorter than that of the Hall of Prayer for Good Harvest to deliver a fullness and uncompromisingness of the main building that features the beauty of ascendance. The "traditional Chinese curved roof" extends smoothly outward with an originally downward eave trajectory curving upward in lifting gesture. Including the green glazed roof tiles for the auditorium top, scarlet colonnades and white balustrades, the color scheme amplifying the confidence and aspiration of New China is colorful yet not vulgar, contrastive and energetic, instead (Fig. 2-115).

Conspicuously, the construction work had imposed a challenge. The contractor, lacking heavy hoist and crane machineries, used thousands moso bamboos, wood boards to build scaffolds to rig up 36 double-meshed angle steel trusses of 1m high, over 280 tons weight and tied with 40,000 rivets for the conical roof to the supporting ferroconcrete columns. The difficulty and massive work can be imagined (Fig. 2-116, Fig. 2-117).

The overemphasis on the architectural profile had its drawbacks. The acoustic effect degraded as the reverberation time increased. The over-extended "oversail eaves" caused the cracking of its supporting cantilevers, and coarse decoration reduced the artistic value of the building. After recent renovation, these deficiencies had been removed. In general, those defects did not belittle the merits. The Chongqing Great Hall of the People is an important pioneering work of the "architectural integration of Chinese and Western" during the New China era.

Case 2 Cultural Palace of Nationalities, Beijing

Located at the west of Chang'an Jie of Beijing, the Cultural Palace of Nationalities was built in September 1959. It is one of first group of "Beijing Ten Great Buildings" (Fig. 2-118). Its construction area is 32,000 square meters. The main building has 13 stories in 67m height, embraced by the east and west annex buildings.

Fig. 2-119 Square pavilion with a double-layered pyramidal roof on the tower of the Cultural Palace of Nationalities, Beijing

Fig. 2-120 Exterior of the tower in the Cultural Palace of Nationalities, Beijing

Fig. 2-121 Corner handling of annexes in the Cultural Palace of Nationalities, Beijing

Fig. 2-122 Details of entrance and exit at the first floor of the Cultural Palace of Nationalities, Beijing

The central exhibition hall, extending northward with its roof with curved oversail eaves covered by peacock blue roof tiles and white towering body, is demure, unconventional and magnificent, emitting unique artistic appeal of the "architectural integration of Chinese and Western".

In 1951, Chairman Mao personally suggested building a cultural palace of nationalities to signify the great union of all nationalities and to provide an activities center for ethnic minorities. The suggestion was not executed until 1954. The renowned architect Zhang Bo was appointed to preside over the design. The original idea was to have a "traditional Chinese curved roof" to represent the national characteristics. However, at that time, the "Three-Anti Campaign" against three evil deeds of "corruption, waste and bureaucracy" was at its boiling point, the original idea was thought profligate by architects. It was endorsed by the former Premier Zhou Enlai during his review and finally got implemented.

The building of the Cultural Palace of Nationalities has its historical significance of carrying forward national inheritance and bring forth prospect. Except the high rise of the Headquarters of the Bank of China in Shanghai built in 1937 introduced "Chinese classical style", there was not any other successful precedents until the early 1950s. This imposed creativity and courage challenge to the architect. Therefore, from the political aspect, its implication weighed more than that of the edifice of the Bank of China, while New China urged to establish its political, cultural and architectural paradigm that anchors the positioning of specifics. Besides the historical complexity of domestic multi-nationality affairs and the new social situation of that time (as one of the original purposes was to have an accommodation place for Dalai Lama and Panchen Lama in Beijing), the psychological reaction of other ethnic groups was also a consideration factor. The

Fig. 2-123 Exterior of Beijing Railway Station

Fig. 2-124 Rendering design Plan of Beijing Railway Station

Fig. 2-125 Main hall of Beijing Railway Station in a ferroconcrete double-curvature arch span structure

balance achieved by the architect cannot be overstated.

Adding a square pavilion with double-layer eaves pyramidal roof is a nice design touch to bring the building in command of the entire site (Fig. 2-119). Parallactically, the elevation plan (from drawings) displays an uneven towering proportion, yet when zooming in the perspective, the view shows balanced aspects. This reveals the profound visual effect manipulation and sophisticated design techniques of the architect.

In addition, the shrewd window design on the elevation plan of the main building delivers the needed natural lighting. The continuous window awning on the front façade is different from any previous buildings. The sprightly tempo and delicate molding bring out solemnity and geometry. The building with varying parallax emits a watchtower charm of Tibet (Fig. 2-120 to Fig. 2-122).

The roof and window awning covered with peacock blue glazed roof tiles present the aura of ethnic minorities, which was unprecedented in the palatial style of the Ming and Qing dynasties. This again reveals the architect's political wits in expressing the sentiment of ethnic minorities and embracing the national unification. Obviously, without proper materials this task could not be accomplished. At that time, domestically, most glazed tiles were subdued and heavy colored in yellow and green, with some in azure. Zhang Ximing, the principal of the construction project, visited Jiangsu Yixing Glazed Tile Factory and an old craftsman located a sample of peacock blue tile, for which it was the exact style Zhang was looking. Zhang had the sample analyzed and produced in batch according to the formula. The batch was used on the building to create an architectural masterpiece.

Case 3 Beijing Railway Station

Beijing Railway Station (in brief, Beijing Station) is at 13 Maojiawan Hutongjia, Dongcheng District. It is in

Fig. 2-126 Design of main entrance to Beijing Railway Station adopting the modeling elements of folk dwellings

Fig. 2-127 Nanjing Xiaguan Railway Station designed by Yang Tingbao

Fig. 2-128 A grand, bright space delivered by a double-curvature arch span structure in the main hall of Beijing Railway Station

Fig. 2-129 A grand, bright space created by large areas of glass windows in the side hall of Beijing Railway Station

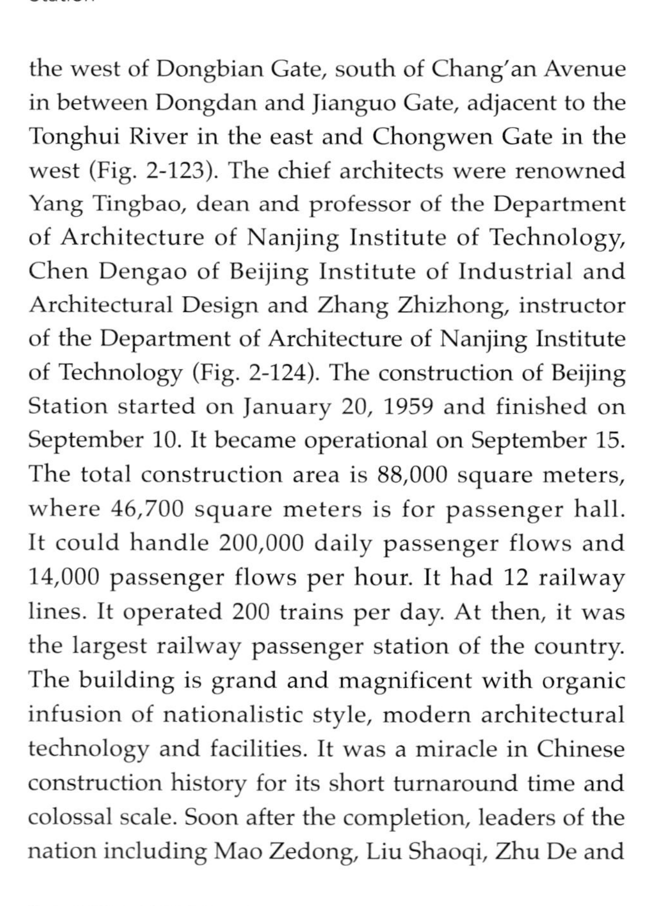

the west of Dongbian Gate, south of Chang'an Avenue in between Dongdan and Jianguo Gate, adjacent to the Tonghui River in the east and Chongwen Gate in the west (Fig. 2-123). The chief architects were renowned Yang Tingbao, dean and professor of the Department of Architecture of Nanjing Institute of Technology, Chen Dengao of Beijing Institute of Industrial and Architectural Design and Zhang Zhizhong, instructor of the Department of Architecture of Nanjing Institute of Technology (Fig. 2-124). The construction of Beijing Station started on January 20, 1959 and finished on September 10. It became operational on September 15. The total construction area is 88,000 square meters, where 46,700 square meters is for passenger hall. It could handle 200,000 daily passenger flows and 14,000 passenger flows per hour. It had 12 railway lines. It operated 200 trains per day. At then, it was the largest railway passenger station of the country. The building is grand and magnificent with organic infusion of nationalistic style, modern architectural technology and facilities. It was a miracle in Chinese construction history for its short turnaround time and colossal scale. Soon after the completion, leaders of the nation including Mao Zedong, Liu Shaoqi, Zhu De and Zhou Enlai paid visits to the station. Chairman Mao inscribed "Beijing Railway Station" in person and the inscription is still in use. The architectural historian Lai Delin remarked, "Beijing Railway Station is the only urban infrastructure construction of 'Beijing Ten Great Buildings' and the only case of integration of the nationalistic style with modern technology".[1]

Beijing Station has two merits in the design art of the "architectural integration of Chinese and Western". First, it is the integration of double-curvature arch span structure in the nationalistic style. The central passenger hall has two 35-meter directional spanning to accommodate the capital's passenger flow, which could be done with other structural designs, for example, trusses or meshed struts. However, the designer chose the novel, structural-wise rational yet simpler double-curved ellipsoidal shape (Fig. 2-125). The two added nationalistic style bell towers capped with pavilions with double-layer eaves pyramidal roofs on the two sides, symmetrically, of the main hall, the portico topped with continuous arches, and the details (Fig. 2-126) resembling dormers and sitting on the top of the eaves mouths of each arch, in some degree, balance the modernistic structure emphasis

1 Yang Yongsheng, & Gu Mengchao (Ed.)(1999), *Chinese Architecture of 20th Century* (pp 241—242), Tianjing: Tianjin Sience & Technology Press

Fig. 2-130 Front view of the main building of Beijing Friendship Hotel

Fig. 2-132 Details of the main entrance to the main hall of Beijing Friendship Hotel from the front

Fig. 2-131 Rear view I of the main building of Beijing Friendship Hotel

Fig. 2-133 Rear view II of the main building of Beijing Friendship Hotel

of the double-curved ellipsoidal span. If having the pavilions and the portico removed from the picture, the building would have been just another modern architecture unrelated to a "nationalistic style" transportation facility. The circumstances of times, location, specific objective, and political sensitiveness required architects to weigh the consideration of a large public building as city's landmark and the nationalistic symbol at capital's portal over the "Structuralism", after experiencing the power of "repudiation" of the building of the Peace Hotel. Or, perhaps, the design of balancing the two preceding factors was successful, as a result of "ideological remolding", that architects had gained their wits and become mature in parrying, deflecting and compromising amid political influences.

The character formation of the large transportation building, especially Beijing Railway Station, did not impose too much difficulty in the architectural design to the experienced Yang Tingbao who over a decade ago designed the expansion project of Nanjing Xiaguan Railway Station (Fig. 2-127). Although time had changed, the architectural character formation can still be learned. The large window curtains accompanying arches under the double-curved ellipsoidal span not only provide natural lighting but also radiate vivaciously features of the transportation portal, which

Fig. 2-134 Details of the secondary entrance to the main building of Beijing Friendship Hotel from behind

Fig. 2-135 Peace dove shaped chiwen on the roof of the main building in Beijing Friendship Hotel

is simple, lucid, open, bright, spatial and towering. The design is touched with convenient portico, large glass curtains and waiting halls on two sides supported by tall and slender colonnades (Fig. 2-128, Fig. 2-129), which compared with Nanjing Xiaguan Railway Station, has resemblance in styling vocabulary sacrifice, such as the inclination of space created by using shallower arches and tall columns.

In June 2008, to welcome the Olympic Games, Beijing Railway Station added English signboard "Beijing Railway Station". Although the Beijing West Railway Station and Beijing South Railway Station are operational, trains connecting the rapid developing East China, reinvigorating industrial base of Northeast China and conveying tourist trains in supporting the growing "tourism economy" are all started from here with regular passenger flow volume. The over five decades old Beijing Railway Station still emits the same capital glamour of China and Beijing with its unfaltering transportation hub position, unimpressed by newly added attractions of "Bird's Nest" and "Water Cube" and preserving the nationalistic style with its integrated architectural "Chinese elements".

Case 4 Beijing Friendship Hotel

Beijing Friendship Hotel is at 1, Zhongguancun South Street of Haidian District in the core zone of Zhongguancun High-Tech Park of Beijing, abutting higher educational institutes, such as the Peking University and the Tsinghua University. Its plot area is 335,000 square meters of which over 200,000 square meters is allocated as green zones. The site has a beautiful landscape and scenery. The building is simple yet elegant fully embracing China's national characteristics. The main building, South Building and North Building were built in 1954, with funding of 100 billion yuan (10 million RMB, The first series of Renminbi banknotes was introduced during the Chinese Civil War by the newly founded People's Bank of China on December 1, 1948. This series also called "Old Currency", which 10,000 yuan is equal to 1 yuan of the 2nd series and later called "New Currency") allocated from the state revenue approved by the State Council. The structure is in ferroconcrete and mixed

Fig. 2-136 Peach dove-themed deity and beast statuettes on the roof of the main building in Beijing Friendship Hotel

frame, 4—5 stories in height. The construction area is more than 180,000 square meters. The renowned architect Zhang Bo designed the work (Fig. 2-130, Fig. 2-131).

In the early years of the People's Republic of China, many Soviet experts came to China to assist the economic recovery and construction. Their housing must have been arranged. Therefore, the construction of Beijing Friendship Hotel started in 1953.[1] To let these expatriates enjoy an exotic appeal of a foreign land, the "nationalistic style" had another opportunity to flaunt its glamour. On the top of the front façade, double-layer hip-and-gable eaves cap the building to cover the elevator machine room and fire-protection facilities. On its two sides, there are two pavilions with round-ridge single-eaves roof. The five-story main building and the two side buildings with pavilions on the top are connected by flat-roof tops and pergolas. On the upper portion of the front façade of the main entrance, balconies with balustrades and railing panels in classical design of newels and handrails are added (Fig. 2-132 to Fig. 2-134). Furthermore, the detailed

1 Yang Yongsheng, & Gu Mengchao (Ed.)(1999), *Chinese Architecture of 20th Century* (pp 218), Tianjing: Tianjin Sience & Technology Press

decorations, including the abstracted and modified traditional deities, beasts and other zoomorphic ornaments, create a brand new genre of detail patterns—the palatial dragon-shaped architectural decorative element *zhengwen* (or *chiwen*, main ridge decoration) of the Ming and Qing dynasties are replaced by the "peace dove" (Fig. 2-135, Fig. 2-136). The reason for doing so is quite worth pondering.

Compared with similar elements, including "jade and white silk ornament" of wishing for peaceful coexistence among collectives or personal-collectives in the extant traditional Chinese architectural vocabulary, the "World Peace" object reveals significant divergence in ideology and discourse system based on different zeitgeists and in diverse regional cultures under the international political confrontation. Strictly, before the birth of modern national states, the traditional Chinese culture does not have any cultural element that represents the peaceful relationship between countries, for the "land of heavens" has been the central ideology of Chinese since the Western Zhou Dynasty (ca. 11th century—771 BC) that all things under the sky are the emperor's subjects and lands. Despite being divided into different vassal states in that time, China still remained a unified country in the name of the emperor where harmony was the norm and war was the chaos. The Son of Heavens owns four seas (meaning the whole land under heavens). There is no equal enemy so there is no symbol representing peace between countries. The peace symbol of dove and olive branch is the allusion from *Genesis* of the *Old Testament* of the *Bible* in connection with the biblical story Noah and the Flood. Tracing back the source, it is a typical western cultural element. In today's view, the integration of the peace dove with the contemporarily styled Chinese classical architecture can be considered a "traversing", whose symbolic function is innovative. Conspicuously, the artistic work of the world peace wish had no precedent in the traditional Chinese architecture but was correlated with the international political dynamics. From the splitting of Germany into East Germany and West Germany, Korean War to Indochina Wars, as the Cold War between the east and the west camps was upsurging, New China

Fig. 2-137 Exterior of the north side of the Teaching Building of the original Eastern China Aeronautics Institute (ECAI), Nanjing

Fig. 2-138 Exterior of the south side of the Teaching Building of the original Eastern China Aeronautics Institute (ECAI), Nanjing

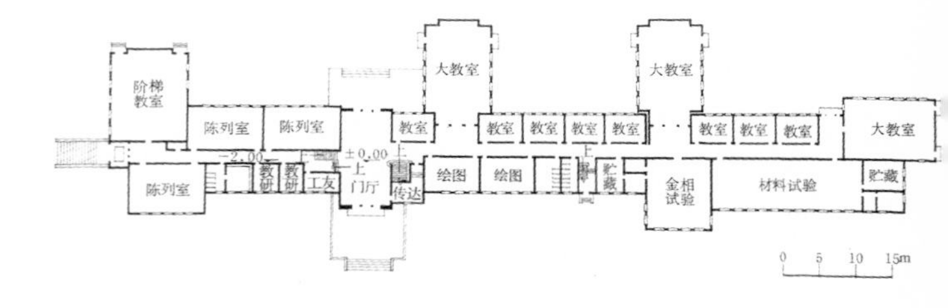

Fig. 2-139 The first floor plan of the Teaching Building of the original Eastern China Aeronautics Institute (ECAI), Nanjing

inescapably was drawn into the long-lasting standoff. However, after centuries warring turbulence, the ancient China and its new regime, just panting a breath, did not have the desire to participate. The entire nation was longing for a peaceful and calm period. Adopting peace dove as the decorative detail for an architectural work serving a diplomatic mission was the end product of the architect's political awareness and personal yearning, also an innovative move revealing the profound background of the times and the great practical significance. However, just as the "traditional Chinese curved roof" in the "architectural integration of Chinese and Western", the architecture arena failed to understand intellectually the bipolar effect of the nationalism, needless to say to comprehend the intertwining historical influence of the "leaning to one side" diplomatic strategy—the assertion to create externally favorable environment by declaring independence and sovereignty, eliminating the century-old impact of the western imperialism and allowing New China to enter the world political stage with a brand new dignity. These gestures exposed the negative effect that the West, facing the international landscape under its own ideology and the scope of US-Soviet confrontation, denied diplomatic relationship with China. Especially, when China was involved into the Korean War, the Sino-US dissension was not avoidable. New China, in its early years, inescapably has been facing suppressions of capitalism led by US in politics, economy, culture, military and other aspects. The "friend or foe" binary ideology narrowed the flexibility of New China's maneuverability on the international political arena. Whereas of this sense, not until the dissolution of the Soviet Union and the fall of communism in East Europe that ended the Cold War in the 1990s of last century, had the "peace dove" to Chinese finally changed from an extravagant wish to a reality. The political vision designers bestowed to the innovative detail of Beijing Friendship Hotel proved to be forty years ahead of the reality.

Case 5 Teaching Building of Eastern China Aeronautics Institute (ECAI), Nanjing

Jointly designed by teachers and students of the Department of Architecture and Department of Civil Engineering of Nanjing University of Aeronautics and Astronautics and teacher and directed by the renowned architect Yang Tingbao in 1953, the Teaching Building is on the campus of Nanjing Agricultural University at today's address, 6 Tongwei Lu of Baixia District of Nanjing City (Fig. 2-137, Fig. 2-138). The main building body has 2—3 levels and a part of it in five levels. Its structure is a ferroconcrete frame. The total construction area is 5,000 square meters.

The original Eastern China Aeronautics Institute (ECAI), Nanjing was formed by merging the

Fig. 2-140 Diversified roof combination of the Teaching Building of the original Eastern China Aeronautics Institute (ECAI), Nanjing

Fig. 2-141 The Hillversum City Hall by Dutch architect Willem Dudock in the 1930s

Fig. 2-142 Details of the tower of the Teaching Building of the original Eastern China Aeronautics Institute (ECAI) ,Nanjing

Department of Aerospace Engineering of Jiaotong University, Nanjing University (former National Central University) and Zhejiang University under the reinstitution effort of higher educational institutes in 1952 after the founding of New China. In late 1950s, because of the international environment and domestic construction needs, the University was moved in stages to Xi'an, and the then campus became Nanjing Agricultural College (today's Nanjing Agricultural University) ever since.

Different from Beijing's political sensitivity and drone, architectural activities in Nanjing in the early stage of New China was quiet and mellow. The Teaching Building was built under this background. Restrained by funding and scope, the architect was forced to adapt the functional performance to the natural terrain in plan layout. The first level foundation used three elevations to reduce earthwork (Fig. 2-139). The overall layout was asymmetrical to give the rarity of dynamic balancing effect seen in the "contemporarily styled Chinese classical architecture". The smaller size yet flexible profile of the "traditional Chinese curved roof" was used to create an innovative and vigorous artistic effect. The design of the main entrance façade is the traditional Chinese three-gate four columns memorial arch with modifications. The lofty staircase is topped by a cruciform ridge of the hip-and-gable roof with double-layer eaves. The rest roofs are large flat root decorated with straight ridges (Fig. 2-140). If ignoring the visual effect of the "traditional Chinese curved roof", it is easy to find the interaction between building blocks revealing the Cubism connotation of early modernism, similar to Hillversum City Hall (Fig. 2-141) designed by Dutch architect Willem Dudock in 1930. If most of the buildings of the "architectural integration of Chinese and Western" mentioned in the preceding are symmetrical to the central axis, lacking initiative and being overcautious in profile and compromising and conservative in design techniques, the Teaching Building of the same "architectural integration of Chinese and Western" inclines more toward the building-block permutation of the early modernism that is liberal, stretching, vigorous and active. More interestingly, the flag mast on the top of the cruciform ridge of the hip-and-gable roof with double-layer eaves is tipped with a red pentagram in noticeable size demonstrating the designer's political wisdom and the characteristics of political formality and symbolism in Chinese architectural culture. The red pentagram on the top of the bell tower of Nanjing University is not alone (Fig. 2-142).

2.6 "Cold War" Taiwan: "Chinese Cultural Renaissance" and Architecture

After restituting its regime in Taiwan and the global political landscape shifted to "Cold War", the Kuomintang reestablished itself amid the Korean War and started the cross-straits standoff for decades. In the first half of the period, or before the early of 1970s, the international policy of the Kuomintang regime was to form alliances with US and Japan against the Soviet Union and the government of the People's Republic of China and to use the seat in the Security Council of the United Nations to empower its defense. The domestic policy was to conduct political reform of the Kuomintang Party, the economic reform of land administration and private economic and infrastructure construction. As the military confrontation being fallen into despair, its focus turned into cultural construction, especially, in contrast with the "Great Proletarian

Fig. 2-143 Exterior of Taipei National Sun Yat-sen Memorial Hall

Fig. 2-144 Details of a portico of Taipei National Sun Yat-sen Memorial Hall

Fig. 2-145 A portrait of Wang Dahong

Cultural Revolution" launched by the Communist Party of China in 1966, it began "Chinese Cultural Renaissance" in Taiwan as a commensuration. Its background ideal was to emphasize the significance of Confucianism in the traditional Chinese culture and recount its significance in contemporary reality to defend its orthodox position and regiminal legality. It reacted against a complete Occidentalization. Furthermore, it urged to "subsume inheritance and embrace innovation, root on moderation and ward off heterodoxy"—to stand fast the tradition and espouse foreign positives. It proclaimed the idea of "absorbing

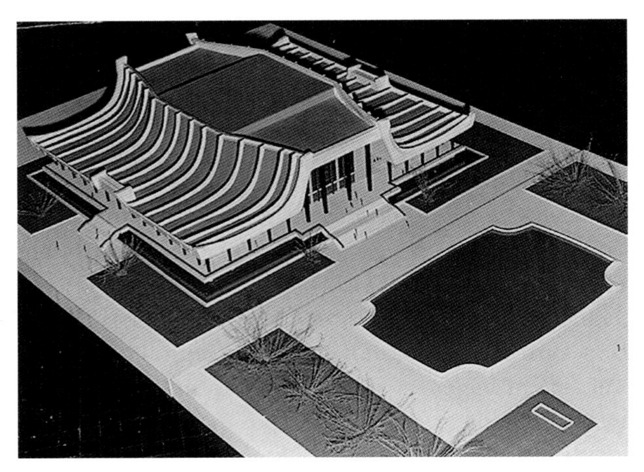

Fig. 2-146 Design model of Taipei National Sun Yat-sen Memorial Hall that won the open tender by Wang Dahong

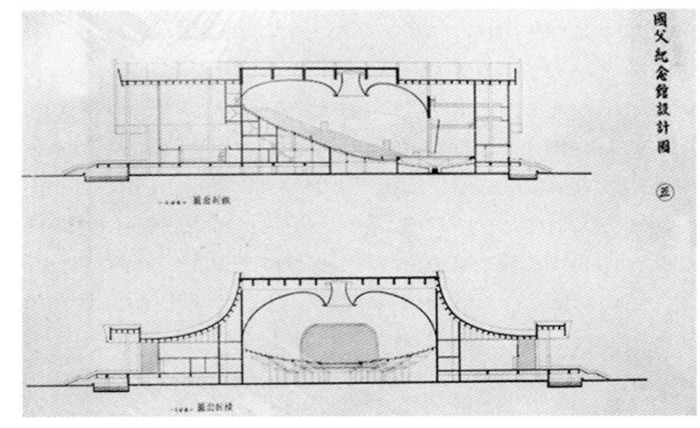

Fig. 2-147 Profile of Taipei National Sun Yat-sen Memorial Hall design that won the open tender by Wang Dahong

Fig. 2-148 Outdoor activities at Taipei National Sun Yat-sen Memorial Hall: chess players and their onlookers

Fig. 2-149 Outdoor activities at Taipei National Sun Yat-sen Memorial Hall: having a rest and viewing

the essence of Chinese and Western cultures and creating a new third culture", which as manifested onto architecture, it endorsed and advocated "Chinese palace architecture". Under this policy, a series of large public buildings of political memorial buildings, science and cultural buildings, hotels and administration buildings were constructed in the format of the "architectural integration of Chinese and Western", such as the "National Sun Yat-sen Memorial Hall", the Grand Hotel, and the Chinese Culture University in Taipei.

Case 1 Taipei National Sun Yat-sen Memorial Hall

Located in Zhongshan Park, at 505, Section 4, Jen Ai Road., Taipei, encircled by Chung Hsiao East Road, Jen Ai Road., Kuang Fu South Road., and Yat-sen Road., Taipei National Sun Yat-sen Memorial Hall was built on the centennial anniversary of Dr. Sun Yat-sen to commemorate the China's democratic revolutionary forerunner (Fig. 2-143). Its construction work started in 1964 and finished in 1972. The Hall includes an auditorium accommodating 3,000 seats, lecture room, library and exhibition room. The plan is a square, where the Hall sits at the center and the rest of functional elements are arranged at its peripheral. The outside has a surrounding cloister. The main structure uses steel-reinforced concrete (SRC). The top is covered by golden glazed roof tiles. The main entrance has entasis columns to lift a flat roof top, forming a solemn and grand portico (Fig. 2-144). The building has 30.4 m in height, 100 m in length on each side, and a construction area of 35,000 square meters. The Hall has 14 gray columns on each side to support the roof top of the "traditional Chinese curved roof", delivering an impressive grandeur of uncompromising perseverance and simple exquisiteness to annotate the ethos of Dr. Sun Yat-sen from the designing architect.[1]

The Hall was designed by the known architect of "Taiwan's modern architectural movement pioneer" Wang Dahong (born 1918) (Fig. 2-145). It is Wang's great feat that won him the open tender on November 6, 1965 [2] (Fig. 2-146, Fig. 2-147), that his wide spanning ferroconcrete structure demonstrating the extraordinary mechanical performance and the beauty of the "traditional Chinese curved roof" in integration received high praise in Taiwan's architecture arena.

1 Xu Mingsong (2007.15), *Wang Dahong— Unremitting Poet of Architecture*, Taipei: Muma Culture

2 Liang Minggang, Zeng Guangzong, Jiang Yajun, & Xie Mingda (2009), In Xu Mingsong (Ed.), *Untold Story of Dr. Sun Yat-sen Memorial Hall--Wang Dahong's Compromise and Tribulations* (pp 2-3), Taipei: Sun Yat-sen Memorial Hall

However, the implementation, per request of the contracting party, was modified to retain only the "traditional Chinese curved roof" even with which that was still a successful annotation of the classical genre by a modernist and was embraced by the public. In 2009, Wang, at 91, received the 13th Taiwan Cultural and Arts Award. His known works, besides "Taipei National Sun Yat-sen Memorial Hall", also include Lin Yutang Residence and the Student Center I of National Taiwan University. The judge of the review panel unanimously agreed that although Wang educated under the western educational system consistently demonstrated his self-awareness of devotion to the integration of the traditional Chinese architecture and the western modernism. "Taipei National Sun Yat-sen Memorial Hall", after 40 years since its construction, is still recognized as one of the outstanding public architecture in Taiwan modern architectural history by the general populace and the industry. Other critics previously considering, "Wang's Chinese minimalism influence brings forward the advancement of Taiwan architectural modernization", was finally confirmed by this award. To the public's consensus, the building provides functions besides its original commemoration and assembly of a public place for recreation and outdoor activities for urbanites (Fig. 2-148, Fig. 2-149), where Fig. 2-143 shows a child is flying a kite on the front square, and urbanites in groups are strolling or sitting along the surrounding cloister to relax in Fig. 2-148 and Fig. 2-149. These reflect one of the possible core values of modernization.

Fig. 2-150 The Grand Hotel in Taipei surrounded by the well-tended plants (from Guan Hua, Nanjing University)

Fig. 2-151 Interior of the Grand Hotel lobby (from Guan Hua, Nanjing University)

During the design process of the Hall, the interaction between the architect and the competent authority was most interesting. After Wang received his award, the design model was presented to Chiang Kai-shek by Wang in person at the "Office of the President, Republic of China". Wang was convinced by the representativeness of modern architecture of Chinese culture of his design, not expecting Chiang politely commented the design to be in "Western flavor". Moreover, Chiang sent Presidential Secretary General Zhang Qun and Premier of Executive Yuan Yen Chia-kan to deliver his instructions to Wang that "the exterior appearance should emphasize Chinese architectural style" and to hand Wang the picture of the Hall of Supreme Harmony of the Forbidden City to follow the idea and redesign a building in "Chinese palace architectural style", which was inconceivable to the aspiring architect! Wang then responded without compromise by saying "One of the missions of Sun Yat-sen's revolution is to overthrow the Qing court epitomized by the palatial architecture, while, now, to use this architecture to commemorate him will be a conspicuous grotesque..." Chiang accepted and changed his mind.[1] However, it is obvious to the

1 Xu Mingsong (2007.15), *Wang Dahong— Unremitting Poet of Architecture*, Taipei: Muma Culture

known that the architect's successful persuasion of the very authority at then was the strategy of perpetrating a practical manipulation of the core perception, because the Qing court is incomparable to "Chinese palace architecture"! Although "Chinese palace architecture" is the official architecture of North China during the Ming and Qing era, the palace architecture was inherited from the imperial architecture of Han ethnic of the Ming. In other words, the palace architecture was not unique to the Qing Dynasty. It could also be a symbol of the Ming Dynasty. The mobilization mechanism of Xinhai Revolution 1911 could not exclude the complicated contradiction between ethnic groups and the revolutionary slogan —"expelling the Tatar barbarians, reviving China, founding a republic, and distributing land equality among the people" that would easily mislead to the perception of a national revolution amid the rise of nationalism of the Han to drive out ethnic Manchurians and to restore their hegemony. Wang's contest was to raise the contravention to the palace architecture. Untowardly, the question could have been rephrased to identify the inexpedience of using Ming's architecture to commemorate Sun Yat-sen. The insolvency is that the revolution let by Sun Yat-sen was to topple the feudal monarchy represented by the Qing court and not limited to that the palace architecture has the equal implication. The Ming Dynasty ruled by the Han Chinese was also a feudal monarchy that its architecture should not be used as a reference. It is perceptible that Wang circumvented the feudal monarchy by evasion.

Fig. 2-152 Exterior of the guestroom balcony in the Grand Hotel, Taipei (from Guan Hua, Nanjing University)

Fig. 2-153 Interior of the guestroom balcony in the Grand Hotel, Taipei (from Guan Hua, Nanjing University)

Case 2 Grand Hotel in Taipei

The Grand Hotel in Taipei, one of the ten great hotels of the world, a landmark of Taipei, adjacent to the lakeside of the Keelung River, a scenic spot of Taipei and Taiwan (Fig. 2-150), is situated southwest of Jiantan Mountain of Taipei. In this book, the one introduced was built as an expansion project in 1971. It is 14 stories high in "Chinese palace architecture". Regardless of the time and perspective, the carved beams and columns, upward curving eaves and eaves corners and golden glazed roof tiles radiate elegance and palatial regality in unique Chinese vertu. The unusual development history of Taiwan adds mysterious veil to the building. The Grand Hotel was constructed in 1952, sponsored by the "Duen-Mou Foundation of Taiwan" chaired by Soong May-ling. In the 1950s of last century, due to lack of five-star hotels to accommodate foreign dignitary guests, the construction of a "National Guest House" was discussed. Soong selected the Taiwan Hotel site because of its geomantic auspiciousness in the vicinity of the Keelung River and the hillside of Yuanshan. Although, Taiwan Hotel showed its ages but the surroundings offer beautiful scenery and serenity with panoramic view of Taipei City. Soong particularly liked to entertain her guests here. The building determined by Chiang Kai-shek was "Chinese palace architecture" style to propagate Chinese culture (Fig. 2-151). Architect Yang Cho-cheng was appointed to preside over the project.

Fig. 2-154 Exterior of the Hall of Great Benevolence of the Chinese Culture University in Taipei (from Guan Hua, Nanjing University)

Fig. 2-155 A portrait of Lu Yujun

The unique feature of the Grand Hotel is the breakthrough of using the usual integration of the "traditional Chinese curved roof" with high rises and not opening windows directly on the exterior walls but to have continuous cloister-type balconies at each story in concert with the functional needs of the hotel. The approach not only blocks the stern sunlight in the Tropic of Cancer zone but also provides a relaxed open view, forming a spatial effect for guestrooms. The exterior structural elements of continuous balconies and surrounding parapets are virtuosically and exaggeratedly rendered in Chinese classical architectural style columns, beams and architraves in unusual scale. The balustrades and corbels (or, *queti*) between columns and architraves of levels create dynamic shades, to wit "gray space" (Fig. 2-152, Fig. 2-153), providing spatial and ambivalent beauty different from the heaviness and stiffness of those of Shanghai Bank of China and the Cultural Palace of Nationalities, Beijing. The double-layer hip-and-gable "traditional Chinese curved roof" adds the grand and magnificent glamour of Chinese official building, balancing the extravagance of the "traditional Chinese curved roof" in the sense of a structure-wise rationale that confirms the success of the artistically handling the special latticed balcony design by "room" and "level". From the effect it creates, as a landmark, the designer's feat is recognized without reservation.

Case 3 Hall of Great Benevolence of the Chinese Culture University in Taipei

The Hall of Great Benevolence of Taipei Chinese Culture University was built in 1965. Today, it is used by the College of Arts and the College of Foreign Languages. It is also called "Zhihui Pavilion" (Fig. 2-154). Its structure is ferroconcrete, in six stories high. The designer was the known architect Lu Yujun (Fig. 2-155).

Chinese Culture University was established in 1961 by the known geographer, historian, and the initiator of the contemporary anthropogeography master Chang Chi-yun. It is a private university at the foothill of Yangmingshan Mountain in the Shilin District of Taipei, adjacent to Yangmingshan National Park. In the beginning of its preparatory stage, Chang, as the incumbent minister of Department of Education decided to have "Chinese Culture Research Institute" founded first, which was later renamed to "Far East University" and changed to "College of Chinese Culture" by Chiang Kai-shek in person with the interpretation of "Far East" as a Euro-US geographical nomenclature. "Chinese Culture" as its name has profound implication, and it has become the basis for the development and the university's motto. Specifically, the essence of the traditional Chinese culture is in its literature, history and philosophy, while, the contemporary occidental civilization is in its science and democratic political system. The renaissance of Chinese culture and propagation must "carry forward the orthodoxies of Chinese and the Occident with the cohesion of the essences of both". Therefore, the base education of the university must be comprehensive. As becoming a comprehensive university, it balances the development of liberal arts, social science and science and technology, as well as the basic tone of the campus planning and architectural design.

Lu Yujun (1904—1975) studied on a work-study basis in France in 1920. Later, he entered the École Polytechnique (commonly known as Polytechnique). He worked as a researcher in the Institute of Urban Planning (Institut d'Urbanisme) of Université de Paris) in 1925. He returned to China in 1929 and served at the Examination Yuan of Nanjing National Government. In 1949, he moved to Taiwan and established the Department of Architecture and Urban Planning of Chinese Cultural University in 1961. Compared with same age scholars, Lu was a highly productive architect and scholar. In the early 1930s, he planned and designed the Examination Yuan, Examination Committee, Grand Examination Hall and new building for the Ministry of Civil Service, and presided over the construction of the auditorium of National Central University. One of his representative works after relocating to Taiwan was the campus planning of Chinese Culture University and the design works

Fig. 2-157 Exterior of roof deck of the Hall of Great Benevolence in Taipei Chinese Culture University (from Guan Hua, Nanjing University)

Fig. 2-156 Lu Yujun (the fourth from the left) present at the groundbreaking ceremony of the Hall of Great Benevolence in Taipei Chinese Culture University

Fig. 2-158 A corner of the Hall of Great Benevolence in Taipei Chinese Culture University (from Guan Hua, Nanjing University)

Fig. 2-159 The Examination Yuan of Nanjing National Government built in 1930s under the leadership of Lu Yujun

of the "Hall of Great Achievement", "Hall of Great Benevolence", "Hall of Great Rite", and "Hall of Great Mercy" (Fig. 2-156).

The unique feature of the Hall of Great Benevolence is the " 器 "-shaped pavilion building complex (literally *Tool*) with the "traditional Chinese curved roofs" populating the upper hill expanse and delivering a magnificent view. It is unusual to find such a courageous attempt in the "contemporarily styled Chinese classical architecture" group of the same period, especially, the center building has an octagonal pyramidal roof, surrounded symmetrically by smaller buildings at four corners with square pyramidal roof with double-layered eaves. There are nine roofs in total in the same style (Fig. 2-157, Fig. 2-158). This approach apparently is quite different from the typical method of using one large-scale roof to lead the cluster. Lu's design of the five halls, including his other works, such as Taipei National Science Museum, were criticized as merely a formality stacking. The Hall of Great Benevolence, relatively, is more integrated. However, Lu's presentation of Chinese Culture University, compared with his early work Nanjing Examination Yuan (Fig. 2-159), is "vigorous" and "intense", which should be related to the background of a particular time.

In Chinese contemporary and modern architecture history, as one of the pioneers who introduced modern architecture into China, Lu devoted to the exploration of both the theory and practice, for the frequent disconnection between the two. During his study in France, it was the most active moment of the architectural rationalism led by Le Corbusier (1887—1965, also known as Charles Edouard Jeannert-Gris) and the upswing to the prime of the architectural modernism in Europe that various architectural languages and urban planning emerged. The influence of these dynamics had on Lu cannot be overstated.

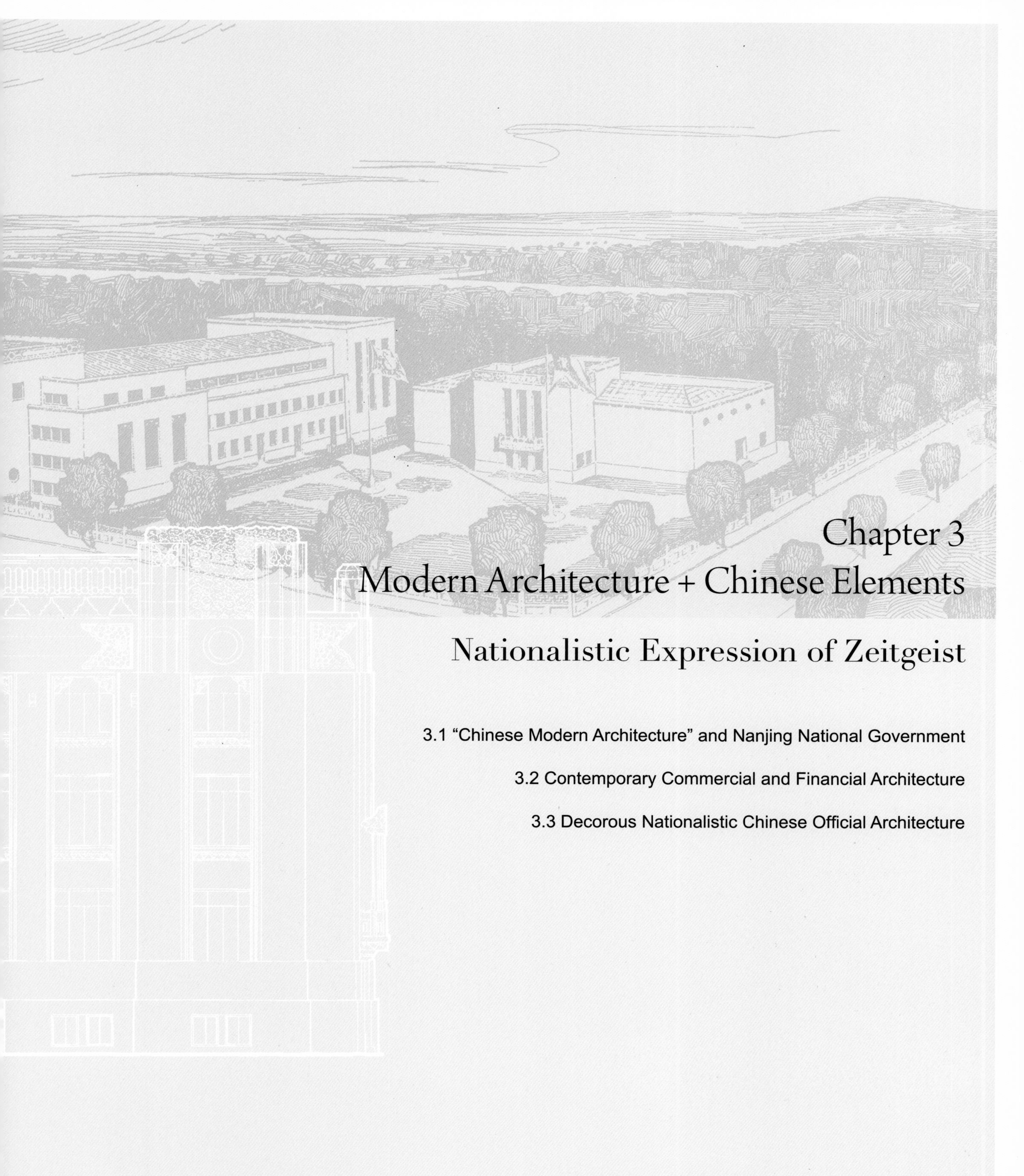

Chapter 3
Modern Architecture + Chinese Elements

Nationalistic Expression of Zeitgeist

Following the founding of the Republic of China in 1911, the rising nationalism ideology had been elaborated into an inviolable group consciousness directing the socio-ascendant discourse and Zeitgeist with the inclusion of ethnic diversity in forming a modern "nationalistic" awareness that sparked off the emergence of the ideological expression symbols of the "rejuvenation of the Chinese nation" and technological vocabulary. The awareness once been formed became the mainstream thought through the entire 30s—40s of the 20th century, leading to the multi-directional and multi-tiered exploration of the nationalistic renaissance architecture of which two formats—the neo-nationalistic format and the palatial format—were categorized by their roots from the structural formal logic of Western architecture and the image elements of the ancient Chinese architecture (Qing's architecture, in particular). The two vary in styles. The palatial style, as a monolithic antiquing art, extracted the entire exterior image elements from the ancient Chinese architecture, especially, the traditional Chinese curved roof, and the neo-nationalistic style, as a contracted antiquing art, reproduced partially the decorative elements of the ancient Chinese architecture, or, epitomized them into abstractions of symbolic images, which by embracing the full-body modern design, embodying dexterously the nationalistic image with conspicuous Zeitgeist and national symbolism, signified the exploratory value of the transformation process of Chinese architecture from the traditional to modern style.

In the early 30s of the 20th century, the advancement of the contemporary Chinese architecture was vigorous and at a pivoting stage. The modern function and technology were driving the materialization of the architectural activities, while the exploration of the art of nationalistic renaissance architecture was at an intersection of China and the West, tradition and modern, or perhaps, was facing the challenge to define the modernity expressions for the essence of Chinese tradition. As to the reality, architects were perplexed by the functional, technological and economic factors in implementing the contracted ancient palatial architecture. Added by the impetus of maturing international modernism that was becoming the mainstream of creation, the heralds of China's contemporary architects came forward following the trend to find novel approaches of integrating the nationalistic renaissance and modernity expression amid the dilemma among the traditional architecture, modern technology, modern functions, and costs associated with the palatial architecture or the traditional Chinese curved roof. Courageously, they formulated a new genre of the architectural integration of Chinese and Western: the "Neo-Nationalistic Style" or "Contracted Antiquing Nationalistic Style". Most were in Western style with flat roof or modern gable roof in ferroconcrete built over a planar layout in modern geometric composition with decorations at specific areas mimicking the traditional details and patterns, such as cornice, sumeru pedestal, metope, tracery windows and porch. The interiors were decorated with flat ceiling panels and color paintings. These ornamental details were used as symbolic representation of nationalistic characteristics not as large componential blocks like the traditional Chinese curved roof. The design, in fact, was an attempt trying to satisfy the needs of new architectural functions and manifest the national traits of Chinese culture at the same time. The style formed the modern nationalistic architecture or a hybrid architecture that opened a new dimension for China's architecture modernization and nationalization.

The rise of Chinese nation, as a national mentality to reestablish the national esteem and carry forth the traditional Chinese culture in the modern world after trouncing the feudalism, had extended into the New China era, which provoked a new round of pursuit of the nationalistic style and historicism in the cultural sphere. Although the ideological fundamental had changed, Chinese architects, entangled between the nationalistic style and the modernism, were still hesitating in defining the creation orientation for the nationalistic style architecture. The perplexity got worsened by the combined factors of economy, politics, new social demands and technological challenge. The uncertainty, therefore, continued.

The approach of adopting the "traditional Chinese curved roof" did not appeal to all architects during the exploration of nationalistic style in similar case as that during the republic era. Many architectural functions or the basic structure were not practical to implement the "traditional Chinese curved roof". Some architects yearned for other different approaches such as adding simple and practical expressions on brick-concrete or ferroconcrete architecture with emphasis on mass and decors with functionalism and Western structure. Meanwhile, they Sinicized the nationalistic style in details, such as decorating the chapiters, adding fascia or *queti* between colonnades, *dougong*-style cornice in brick-and-stone under oversails, Chinese style balustrade, etc, delivering dynamic sensation of Chinese art and Western science in varying perspectives from distant proportion manipulation to near view elaboration. Festoon, coffer, color painting, and embossed painting with golden foils were used in the decoration pattern, which were amiable and making lineage to culture yet in concert with modern function and image. Buildings designed in this way include Beijing Capital Theatre, the Administration Building of Ministry of Construction and Engineering of the People's Republic of China, the National Political Consultative Conference Assembly Hall, Wangfujing Department Store, and Guangzhou Gymnasium, etc. Among them, the Great Hall of the People built in 1959 for the National Day and the Minzu Hotel, with optimized expressions of the western model and the traditional Chinese ornaments, are the representatives of the nationalistic style.

This design integration of Chinese construction elements and decors and modern western structure to express the nationalistic style gained encouragement from the government for its cost-effectiveness. Former Premier Zhou Enlai once instructed that creation must traverse Chinese and Western, and connect the past and current, which means to use whatever serves the purpose. The approach had influenced a long period after 1955. It was also a successful approach after the fading enthusiasm of the traditional Chinese curved roof.

3.1 "Chinese Modern Architecture" and Nanjing National Government

As the capital of the Republic of China, official buildings in Nanjing had definite effect in the national architecture style. The architectural style of the National Government changed from the early imitating traditional Chinese palatial style manifesting the national pride in "China's inherent architectural style" to the cost-effective "neo-nationalistic style" in Chinese decors. In 1930s, Nanjing leading the neo-nationalistic style architecture sprouted a series of masterworks, including Nanjing National Government's Foreign Ministry, the National Central Hospital, Nanjing Central Stadium and the Bandstand of Sun Yat-sen's Mausoleum, which emerged from simple imitating into creation with elaborated touches and became exemplars of the style. However, led by architects, Tong Jun, Yang Tingbao, Fan Wenzhao, Xi Fuquan, et al., who jointly advocated using traditional Chinese decorative patterns at the entrance, coping, and wall base, which was as expedient for national and cultural symbolism as "Chinese modern architecture" was still in cultivation or as an idealism of integrating nationalism and modernity under certain specific ideology? To have a fundamental judgment under the great tension and integration appeal between the social production and formalization of the modern architecture, we cannot reach a conclusive determination. Nevertheless, at least, within a period after 1949, to the exploration of the nationalistic characteristics and modernity, the "nationalistic style" (including the so-called "neo-nationalistic style") still emitted tremendous momentum.

Case 1 Nanjing National Government's Foreign Ministry

The original Nanjing National Government's Foreign Ministry was at today's address 32,

Fig. 3-1 Exterior of the original Nanjing National Government's Foreign Ministry

Zhongshan Beilu, Nanjing, now used by Jiangsu Provincial People's Congress. In July 2001, it was designated as a historical monument and cultural relic under state protection (Fig. 3-1). The building was designed jointly by Zhao Shen, Chen Zhi, and Tong Jun et al. of Allied Architects, one of the important contemporary architectural firms, during 1932—1933 and the construction work was finished in 1935. The design guideline of the building was to lay out and model the building according to modern technology and the building's function, not to copy completely the western style and not to adopt entirely the traditional Chinese palace architecture, but to achieve the "neo-nationalistic style" and reflect the ethos of the time with the traditional Chinese architecture sensation.[1]

Fig. 3-2 Night scene at the entrance to the original Nanjing National Government's Foreign Ministry

The plan layout of the building is in shape of a "T". A portico was built over the entrance (Fig. 3-2). It has one underground level. The central section has five levels above the ground and four levels at its

1 Liu Xianjue (2004), China's Contemporary Architectural Art (pp76), Wuhan: Hubei Education Press

two lateral sides. The entire building plan is done in symmetrical classical western style. The staircase of the central section is the main elevation traffic access that connects the front and the back leg of the "T". The width span is 51 meters; the depth is 55 meters, and the construction area is 5,050 square meters.

The building uses western flat roof in simple and modern geometric expression. All elevation sides have three sections, base, wall and entablature. The base plinth is pasted with cement mortar delivering a sturdy look. The wall is veneered tightly with deep brown Taishan bricks to show its solemnity. Brown glazed bricks are used on the friezes to make low reliefs and simple *dougong* decors. This is a sophisticated antiquing method to express nationalistic taste (Fig. 3-3). The colonnades and purlins of the portico at the entrance have simple connections. The projected ends of purlins are decorated with traditional cloud patterns. To satisfy the owner's demand, scarlet columns are used inside the building. Painting and Qing-style color paintings cover columns, beams, architraves, ceiling and coffers. Interior walls are finished with detailed traditional wall decors. Balustrade, panels, doors and windows are patterned with traditional Chinese designs. The interior design is in a typical antiquing model, not in concert with the contracted antiquing approach. This probably is related to restrictions imposed onto architects by the society and owner (Fig. 3-4 to Fig. 3-7). Architects serve the society and they must face the reality not to indulge themselves in narcissism. Posterity must recount the past architectural activities with the then social development background to correlate and assess properly architects and their works.

Fig. 3-3 Dougong-imitated cornices of the original Nanjing National Government's Foreign Ministry

Fig. 3-4 Lobby of the original Nanjing National Government's Foreign Ministry

The original Nanjing National Government's Foreign Ministry was a valuable experience for the contemporary Chinese architects in searching for a new orientation of architecture development. As an exemplar of neo-nationalistic style, the building echoed the design strategy of the inclusion of "nationalistic traits" and "scientific approach" in satisfying the economic reality and self-value projection of architects. It conveyed significance in the architectural

Fig. 3-5 Interior of the original Nanjing National Government's Foreign Ministry

Fig. 3-6 Beams and ceiling of the original Nanjing National Government's Foreign Ministry

Fig. 3-7 Interior of an office in the original Nanjing National Government's Foreign Ministry

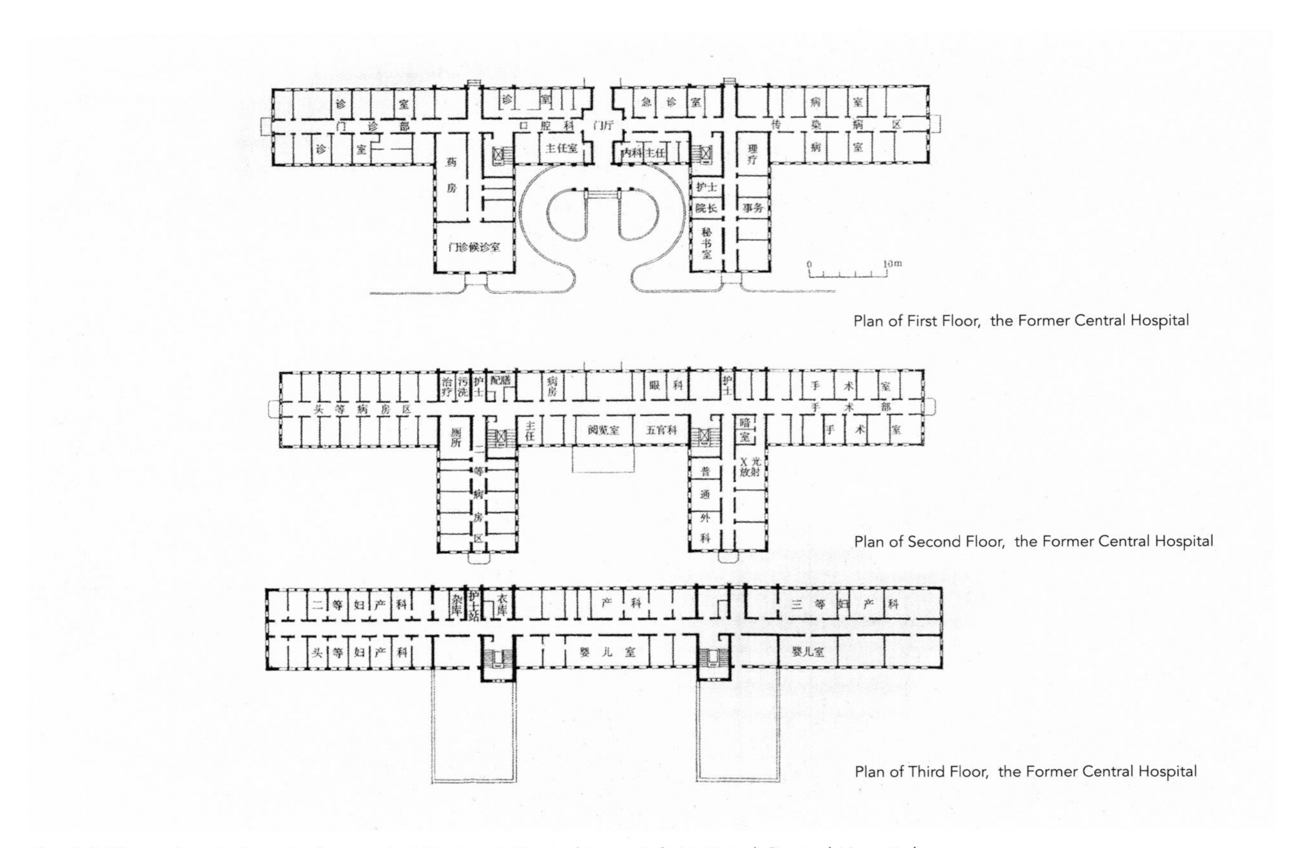

Plan of First Floor, the Former Central Hospital

Plan of Second Floor, the Former Central Hospital

Plan of Third Floor, the Former Central Hospital

Fig. 3-8 Plans of each floor in the original National Central Hospital National Central Hospital

advancement and social assessment, as well as an influence to architects exploring Chinese architecture with nationalistic characteristics.

Case 2 Original National Central Hospital

Established in 1933, located at today's 305, Zhongshan Donglu, Nanjing, the National Central Hospital was the largest and the most comprehensive national hospital in the capital region. It is now Nanjing General Hospital of PLA Nanjing Military Region. The main building as a centralized ward building, arranged symmetrically similar to a ∏ shape is four-story high. The construction area is over 7,000 square meters that is divided by modern functions into out-patient, operation, ward and administrative areas (Fig. 3-8 to Fig. 3-10). The overall layout coincides with peripheral access roads and environment. Its functional division is clear; all traffic routes are planned well, and the space configuration is reasonably done. After over 80 years, the expansion and layout of the hospital still follow the original design plan, revealing the professional competence and vision of the architect.[1] The design leaped out of the then popular revivalism, presenting its "architectural integration of Chinese and Western" characteristics in a different approach.

The architectural model is in a brand new architectural aesthetics, which uses a flat roof and geometric building blocks to deliver simple and lucid western modernism. Its façade composition has three sections in left-right symmetry. The central section has

1 Yang Tingbao (2001), *Architectural Design Collection* (pp 68), Beijing: China Architecture & Building Press

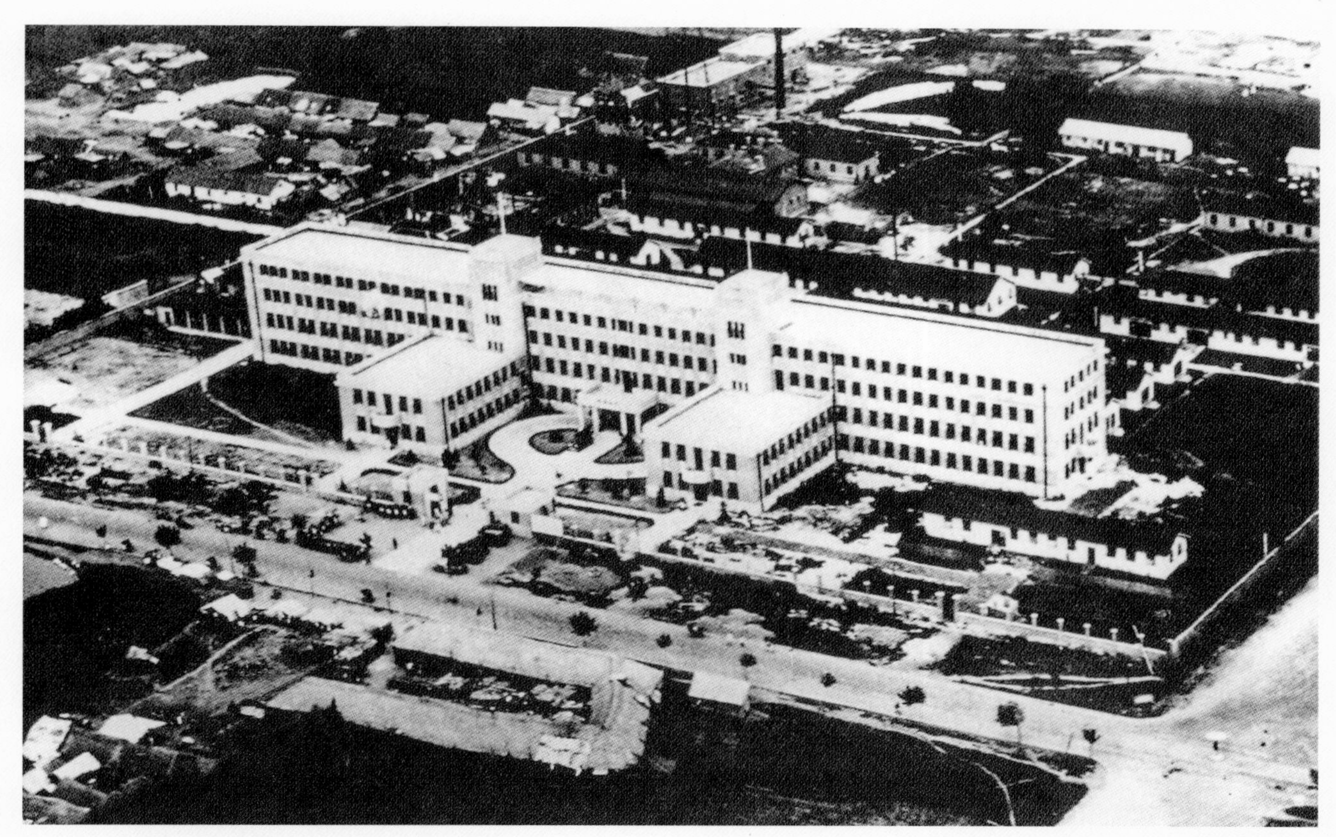
Fig. 3-9 The National Central Hospital in the 1930s

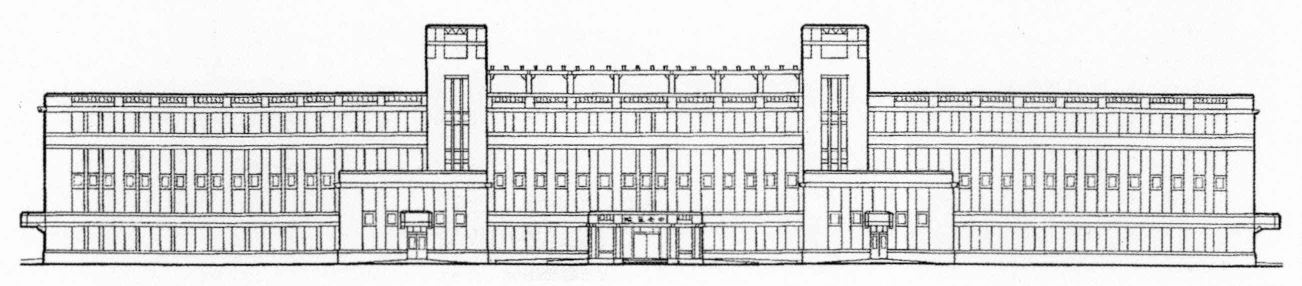
Fig. 3-10 Main elevation of the original National Central Hospital

two staircases (elevators) on the left and right sides. After the front hallway, a portico connected by artificial stone pergola decorated with simplified *queti*. The entire central section is the center of the building body (Fig. 3-11, Fig. 3-12). The exterior wall corners of the staircases are smoothened. The eaves of the staircases and the two wings are decorated with patterns in consistence with those of the entrance. The balusters at the entrance are covered with cloud patterns and surbases at its bottom section (Fig. 3-13 to Fig. 3-15). All elevation faces are in light yellow décor bricks and hard stucco with simple uneven texture. The details imitate the architectural expression of the traditional ornament elements, such as floral designs, architraves, fist-shape beam heads, stylobates and dripping channels, which serve the purpose of introducing the nationalistic style. The portico is in three-bay size,

Fig. 3-11 The current appearance of the original National Central Hospital

Fig. 3-12 Partial exterior of the main building of the original National Central Hospital

Fig. 3-13 Corner handling of the staircases in the original National Central Hospital

Fig. 3-14 Entrance to the original National Central Hospital

Fig. 3-15 Balusters at the entrance to the original National Central Hospital

Fig. 3-16 The portico at the entrance to the main building of the original National Central Hospital

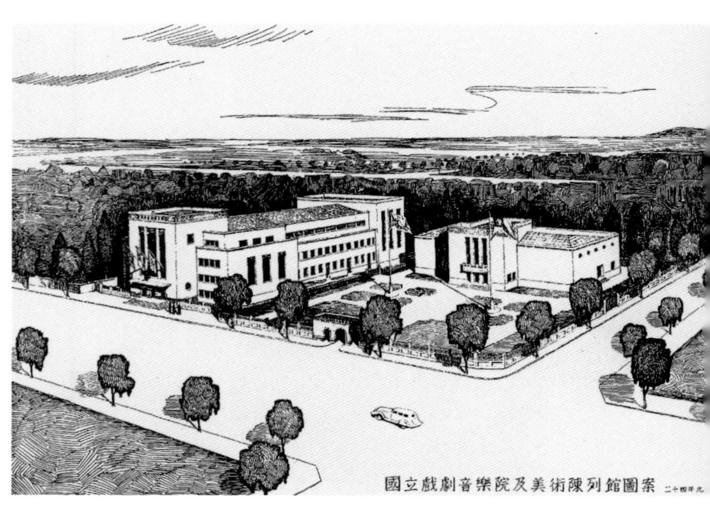

Fig. 3-17 Design drawing of the original National Great Hall of the People and National Art Gallery in Nanjing

emphasized with traditional finish. However, its details abandon the complicated treatment of traditional Chinese architecture. The eaves have protruding fist-shaped beam heads (Fig. 3-16). The entire design is simple and pleasant, retaining a novel and prudent nationalistic style with a presentation of architectural technology, high degree consistency in content and format. The work is an important and outstanding piece in the sprouting stage of Chinese modern architecture.

The chief architect of the National Central Hospital was Yang Tingbao, one of the pioneers and outstanding architects of China's contemporary and modern architecture, whose works were the testimonies of the rise and decline of various architecture genres from revivalism to modernism during China's contemporary reform era. He adapted to, created and innovated in different architecture formats during those transitions. Compared with his early revivalistic work, the influences of Chinese and western architecture are manifested in this building more sophistically not merely straightforward. It is modern yet with cultural ethos elaborated with his mature professional touch.

Case 3 Original National Great Hall of the People and National Art Gallery in Nanjing

The "rise of Chinese nation" was once the ideological expression symbol and the discourse of the contemporary Chinese architecture, while, "Chinese modern architecture", based on the technology of modern architecture, extracted partially the decoration elements from the ancient Chinese architecture and transformed them into abstractions of symbolic images. The following two buildings, although, at some of their parts, using the decoration designs of the ancient Chinese architecture, have their overall styles

Fig. 3-18 The National Great Hall of the People in Nanjing in the 1940s

close to the western modern architecture. They are modern, but have some innovations in the nationalistic style (Fig. 3-17, Fig. 3-18).

The original National Great Hall of the People was the National Theatre of Drama and Music, situated at today's 264, Changjiang Lu, Nanjing, and now Nanjing Great Hall of the People. The building was constructed in the mid 30s of the 20th century. The basic layout is a modern theatre, sitting at the north and facing the south, left and right symmetric, with a four-story main building and one underground level. The Hall has a front lobby, hall, and performing stage. The construction area is 5,100 square meters. The offices are arranged to face the street. The two sides have two two-level refreshing lounges. The interior structure of the Hall is rational with a great acoustic effect design. The main façade is divided into the base, wall and eaves, in typical classical western design. The azimuthal span is also divided into three parts—the towering central block and two sides in flat shape. The entire body is simple. Two-story high glass windows are placed in rows, delivering an amazing Yin-and-Yang rhythmic sensation (Fig. 3-19). The architect adopted the western theater design and the simplicity of modern architectural style, with the elaborated decorations in contracted traditional Chinese patterns (Fig. 3-20, Fig. 3-21). This makes the Hall different from the grand expanse style of the traditional capital buildings and the simplified emulation of the western style in the early contemporary times, making "Chinese

Fig. 3-19 The current appearance of the original National Great Hall of the People in Nanjing

Fig. 3-20 Eaves of the original National Great Hall of the People in Nanjing

Fig. 3-21 A part of the entrance to the original National Great Hall of the People in Nanjing

Fig. 3-22 The current appearance of the National Art Gallery

Fig. 3-23 Entrance to the original National Art Gallery in Nanjing

Fig. 3-24 Eaves and long windows on the main façade of the original National Art Gallery in Nanjing

modern architecture" to stand out. The Hall not only inherited the grand glamour of the traditional Chinese official building but also elaborated elegantly the details, especially in its simple and decorous French style swinging door, view-blocking-free sloping floor, comfortable seating, and excellent hall acoustic effect. It is one of the architectural masterworks during the national government era.

Separated by only a wall, the Hall has the original National Art Gallery in its west. The two buildings have the same style, designed by the well-known architect Xi Fuquan (Fig. 3-22). To be in concert with the grand and solemnity of the National Great Hall of the People, the National Art Gallery emits artistic glamour. The designer used simple structure to have its main building in fluent and simple outline, while expressed the traditional Chinese architectural details on windows, entrance and eaves (Fig. 3-23 to Fig. 3-25). The interior arrangement is exquisite with murals. The exhibition hall is bright and spacious. The overall design is unique with fine Chinese taste. It is a masterpiece of a combination of national art and modern ideology.

When designing these two works, the architect

Fig. 3-25 A stone statue and lamp in the front of the original National Art Gallery in Nanjing

attempted to define the vocabulary for "Chinese Modern Architecture" and abandoned the traditional Chinese classical curved roof"—an integration of modern architectural structure, space arrangement and traditional Chinese architecture elements (such as grasshopper head, the protruding part of a cantilever in 'grasshopper head shape', French-style door 'latticed partition door similar to French-style door' and windows of wooden structural details). In addition, because of the political environment at then, these two buildings of the architectural integration of Chinese and Western were labeled as official buildings. As a neo-nationalistic architectural style of official buildings, their importance cannot be overstated.

Case 4 Original Track Field of Nanjing Central Stadium

In the southeast of Zhongshan Scenic Area in the east suburban of Nanjing, patched in the beautiful landscape, a large sports stadium, Nanjing Central Stadium, built in the 30s of the 20th century, is situated there. The stadium was designed and constructed in ferroconcrete by adapting to the basin terrain. The entire stadium includes track field, swimming pool, baseball field, basketball court, martial arts arena and tennis court. It can accommodate 60,000 spectators. At that time, it was the largest stadium in the Far East. The stadium's plan layout reveals the symmetrical-layout influence of the traditional Chinese commemorative architecture. The track field is placed at the center, and the rest facilities are arranged on its two sides in a balanced formation. The overall layout is grand and vigorous (Fig. 3-26).

The track field is the main building of the stadium and it was the largest track field in then China. It sits on 77 *mu* (about 5.1 hectares) of land in the shape of a rectangle with bilateral semicircles on both ends. The bleachers provide 35,000 seats. There are two entrance gateways, one on the east side and one on the west. The west gateway is also used as the grandstand and the east gateway is a special functional stand (Fig. 3-27, Fig. 3-28). The two gateways are the focus of

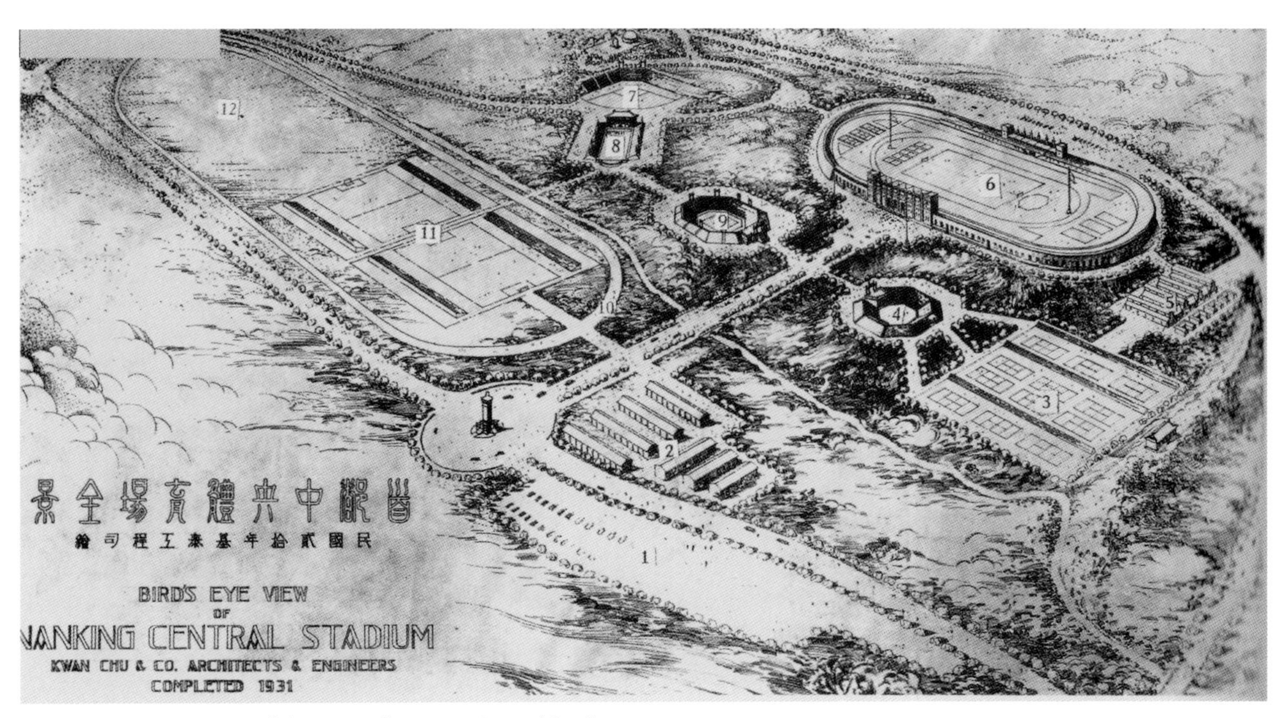

Fig. 3-26 Bird's eye view of the original Nanjing Central Stadium

Fig. 3-27 East gate to the track field of the original Nanjing Central Stadium

Fig. 3-28 The track field of the original Nanjing Central Stadium

the design. The gateway has an azimuthal span of three sections. The middle section has a simple base level and a smooth wall surface. The upper section is in a variant of traditional towering *paifang* with an azimuthal nine-bay span and three-level in elevation, where the combination of columns, architrave and cubic mass appearing to be a *paifang* is actually a design for a grandstand. The top part of the upper section has eight balusters with cloud-patterned capitals and seven small *paifang* rooftops. The stone purlins and architraves decorated with the traditional Chinese pattern details are staggered by bulging and notching to create dynamics. *Paifang*, as a symbol in Chinese culture, was used in ancient days for recognition, memorial, decoration, indication and direction. The architect chose this format to serve those purposes of indication and direction and also solve the transition of spatial sequence. Furthermore, the format is simple and clear, without complicate decorative *dougong*, upward oversail eaves, flying deities and beasts. The design serves as a good example of a breakthrough from the traditional architecture. The two sides of the gateways are balanced and divided into three sections in azimuth and elevation with a longer vertical aspect. The upper sections have reliefs in geometric shape reducing to low-relief cloud patterns as the height increases, matching the cloud patterns with those on the side balusters. This approach creates an innovative classical form. The two gateways have the simple traditional decors integrated with the arch lattice iron gates, revealing the ingenuity of the designer in pursuing the architectural integration of Chinese and Western (Fig. 3-29 to Fig. 3-31).

Absorbing the advanced experience of western modern stadium, the stadium has a rational layout over a generous land lot. In architectural style, namely, a large amount of Chinese architectural elements are included, they are integrated in a guileless form, elegant in simplicity and rich with incorporeal Chinese characteristics that present the whole architecture aggregation in a grand, strong and harmonious integration. It has earned its place in modern architecture history of China, in concordance with a review in architectural trading journal of the same time, "The overall architecture of the stadium, grand and magnificent, applying Chinese architecture

Fig. 3-29 Detail I of the door at the entrance to the original Nanjing Central Stadium

Fig. 3-30 Detail II of the door at the entrance to the original Nanjing Central Stadium

Fig. 3-31 Detail III of the door at the entrance to the original Nanjing Central Stadium

attributes in consistence with the need of the time, awakening the nationalistic awareness of Chinese people to preserve and carry forward Chinese cultural essence, testifies the possibility of bringing innovation into Chinese architectural renaissance movement...".[1]

3.2 Contemporary Commercial and Financial Architecture

There are numerous commercial and financial architecture examples in contemporary Chinese public buildings spreading out in the entire country. They are closely related to the daily lives of the general urban populace. Except those extant stores, the new commercial and financial buildings include banks, department stores, hotels, theaters, clubs, and commercial playgrounds. They, as a new genre, are the largest buildings in the contemporary Chinese urban business district with distinguished architectural art, revealing conspicuously the foreign cultural influence. Most of them have "Western-style façade" meeting the marketing purpose of merchants who by nature have to constantly chase fad and diversity across the classical, compromising or mixing Chinese and Western styles. In addition, this genre usually brought on multilevel, towering or expansive, spanning and high standard buildings, which imposed the requirement on architects to focus on a simpler model to showcase the nationalistic characteristics serving as one factor that led to the creation of the neo-nationalistic style as their representative works.

Case 1 Beijing Bank of Communications

Among many of banks in classical western style, the Bank of Communications at today's Xiheyan outside Qianmen, Beijing, stands out for its architectural integration of Chinese and Western styles (Fig. 3-32).

Fig. 3-32 Exterior I of Beijing Bank of Communications

The Bank built in 1931 was another masterwork of the renowned architect Yang Tingbao. It uses an area around 2,000 square meters in ferroconcrete. The side facing street has four levels. Its bottom level is used as the business hall surrounded by offices and amenities. The underground level is the security vault. The building has clear functional division and reasonable traffic routes. The lobby has raised ceiling design consistent with the modern architectural design (Fig. 3-33, Fig. 3-34). The external sides have straightforward Classicism's design of three sections in both elevation and azimuth, where China's nationalistic style is manifested in details: the bottom section is heavyset and decorous, emphasized by using granite veneers resembling the ancient Chinese palace architectural stylobate; the wall surface is applied with granitic plaster and its upper part has cloud patterns in

1 Chen Xiping (1933), *Architectural Record of Nanjing Central Stadium,* Chinese Architecture (Vol. 1, Issue 3),

Fig. 3-33 Front elevation of Beijing Bank of Communications

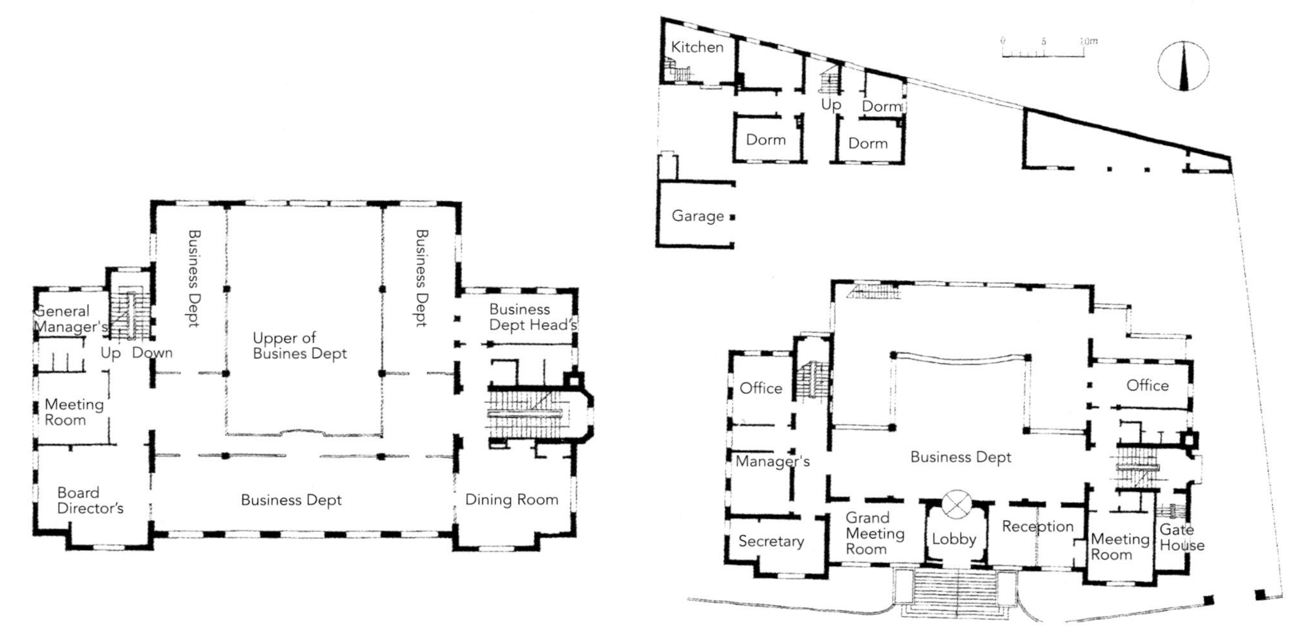

Fig. 3-34 Plans of each floor in Beijing Bank of Communications

chunks; the mouths of oversail eaves are covered with glazed tiles and the underneath cornices are decorated with *dougong*s. The façade facing street preserves the traditional Chinese *paifang* look. Doors are in glass, and windows have glass cover and *queti*. All details are in cloud-pattern stone carving and other festoons to deliver exquisiteness. The entrance gate is festooned and accompanied by two stone lion sculptures (Fig. 3-35 to Fig. 3-37). What is worth mentioning is that the Bank's interior, different from the exterior's abstraction of contracted antiquing style, uses latticed balustrades, color painted beams and architraves and coffered ceilings in full-bodied Chinese classical style, unveiling the owner's true preference (Fig. 3-38).

The design of the Bank was based on Western architecture, not using the traditional Chinese curved roof but incorporated the traditional Chinese decorations and details, which formulated a nationalistic architectural style and matched with then Dashila settings – the traditional business district environment. During the time it was designed and built, the work also emerged into the "neo-nationalistic architectural style" and produced certain influence to the contemporary Chinese architectural design. The contemporary bank architecture illustrated the process of the evolution of the contemporary economic architecture from the traditional to the western format. The Bank is not only recognized as a contemporary

Fig. 3-36 Exterior II of Beijing Bank of Communications

Fig. 3-35 Details of cornices in Beijing Bank of Communications

Fig. 3-37 Ornamental details on the external wall of Beijing Bank of Communications

Fig. 3-38 Ceiling of the business hall in Beijing Bank of Communications

Fig. 3-39 Headquarters of the Bank of China in Shanghai (left) and Sassoon House (right), two eye-catching buildings in the northern section of the Bund

Chinese architecture but also included as one of the landmarks of Beijing City. The Bank became a designated Beijing Cultural Relic Protection Unit in 1995.

Case 2 Shanghai Bank of China

The Headquarters of the Bank of China situated at No. 23, Shanghai Bund is one of the most important tangible cultural heritages of both the banking industry and the financial industries of China. It is a historical monument and cultural relic under state protection. In 1930s, Chinese architect, "seeking for their orientation" amid the world's modernistic design vogue, etched poignant nationalistic awareness. The Bank of China Tower, sitting on Shanghai Bund, in the wake of adopting western architectural style, format and materials, was the only building designed by Chinese architect with the infusion of the nationalistic style, leaping out among the building pack of the "Bund international architecture" (Fig. 3-39).

Shanghai Bank of China Tower, built in 1937, designed jointly by Chinese architect Lu Qianshou and

Fig. 3-40 Interior of the business hall in the Headquarters of the Bank of China in Shanghai

Fig. 3-41 Interior detail I of the Headquarters of the Bank of China in Shanghai

Fig. 3-42 Interior detail II of the Headquarters of the Bank of China in Shanghai

British Palmer & Turner Architects and Surveyors. The tower occupies an area over 3,000 square meters. The east tower built with steel frame structure has 17 levels in 76 meters high. By using the design methodology of functionalism, the designer made rational plan layout according to the purpose of the building and included advanced facilities. The business hall has an area of 1,300 square meters and 10m ceiling setting. At that time, it was claimed to be the largest banking business hall (Fig. 3-40 to Fig. 3-42). The tower's body configuration is in a combination of bulging and notching in elevation, of which the façade of the main tower is facing the Bund and the Huangpu River. The exterior wall uses granite veneers. The overall image is a modern skyscraper. Embellished in the modern architectural expression, traditional Chinese elements are placed at several parts, such as a smoothened traditional Chinese pyramidal spire with a square base that is covered by green glazed roof tiles, stone *dougongs* and lotus patterns on the cornice, traditional perforated windows on the two-side exterior walls from Level 4 to Level 16, and elaborated traditional Chinese

Fig. 3-43 Exterior of the Headquarters of the Bank of China in Shanghai

designs at the entrance and parapets (Fig. 3-43 to Fig. 3-45). Amazingly, the lintel of the front door has a relief of "Confucius Roaming the Warring States". A pair of the sacred beasts of "*Bixie*, exorcising evil spirits" for enriching and protecting the fortune are placed in front of the front door, which once lost or damaged in the turbulent time of history, were restored (Fig. 3-46, Fig. 3-47). Chinese decorative elements are also used in interior, such as cast iron grille patterned with "Climbing Up" and "Eight Immortals" cameo on the ceiling of the business hall. The hybrid of modern and traditional elements reflects the cultural awareness of Chinese architect in their seeking for their orientation. Of course, the eclectic approach of integrating the traditional Chinese architecture and modern skyscraper has its inevitable dilemma, especially, the pyramidal rooftop revealing the unbalanced mass proportion against the skyscraper, although, as a breakthrough in the architectural art, but fails to produce harmonious composition effect, which is inescapably doleful.[1]

Its architectural exploration of the nationalistic style on skyscraper received recognition. For a long period, its style wedding the contracted antiquing

Fig. 3-44 Upper section of the Headquarters of the Bank of China in Shanghai

Fig. 3-45 Annexe to the Headquarters of the Bank of China in Shanghai

Fig. 3-46 Entrance to the Headquarters of the Bank of China in Shanghai and the lintel of the front door

1 Liu Xianjue (2004), China's Contemporary Archtectural Art (pp77), Wuhan: Hubei Education Press

Fig. 3-47 Sacred beast of "Bixie, exorcising evil spirits" in the front of the Headquarters of the Bank of China in Shanghai

presentation and the western decorative style of Sassoon House became the landmark of the northern section of the building complex of the Bund as well as the landmark of Shanghai City, which was used on postal envelops, stamps, postcards and other promotion materials.

Case 3 Shanghai Da Sun Department Store

As the largest and most prosperous trade metropolis, Shanghai, developed rapidly in the construction rush, had a group of western commercial buildings entirely different from the traditional commercial shops erected during 1920s—30s. Those high-rises included the four famous department stores—Sincere, Wing On, Sun Sun and Da Sun, in which Da Sun had the most distinguished features of the time.

Da Sun Department Store, today's Shanghai No. 1 Department Store, located at the bustling

Fig. 3-48 Da Sun Department Store (today's Shanghai No.1 Department Store), located at the bustling downtown of Shanghai

Fig. 3-49 Rendering design plan of Da Sun Department Store

Fig. 3-50 Panoramic perspective of Da Sun Department Store

Fig. 3-51 Pergolas and balustrades on the roof of Da Sun Department Store

Fig. 3-52 Details still with traditional Chinese charm on the interior capitals in Da Sun Department Store

Nanjing Lu of Shanghai, different from other three department stores designed by foreign design houses, was designed by Jitai Construction (or Kwan, Chu and Yang Architects) formed by Guan Songsheng, Zhu Bin (or *Chu Pin*) and Yang Tingbao in 1934 and built in 1936 (Fig. 3-48, Fig. 3-49). The building has 10 stories, a construction area over 28,000 square meters, occupying a land lot over 3,600 square meters in size. The building is in ferroconcrete frame and capped with a beamless top. Its appearance is simple and fluent, clearly revealing its modern functionalism. Its external appearance is divided by its flat rooftop, simple window shapes and walls. The façade only has the traditional Jiangnan (a geographic region in China referring to lands immediately to the south of the lower reaches of the Yangtze River, including the southern part of the Yangtze Delta) style decors on the balustrades on the rooftop and hangers under the pergola to show some Chinese flavor. The face of the first level is veneered with Qingdao black granite. The shop windows are decorated with Chinese black amphibole. The walls use machine-made coal-dust bricks veneered with yellow glazed brick tiles in elevation (Fig. 3-50, Fig. 3-51). The shop floor is paved with Italian marbles and teak floor panels. Da Sun is not only large in size but also equipped with latest facilities. It was the first one offering an underground mall. The market uses the underground and the upper three levels. Office premises, tea house and merchandise displays are set at Level 4. Level 5 is used by the dance hall and pubs. Level 6 to Level 10 are playgrounds. It has seven imported OTIS elevators from US. Most interestingly, it was the first building in China with escalators, which were installed on Level 1 to Level 3 and attracted successfully many spectators. The building has a wide span between colonnades to allow good lighting. HVAC is available on every level (Fig. 3-52).

In 1953, Da sun renamed Shanghai No. 1 Department Store. Before the 80s of the 20th century, it was the largest department store of China, memorized by a few generations for its excited shopping scene.

3.3 Decorous Nationalistic Chinese Official Architecture

After the founding of New China, all the undone needed to be redone. Driven by the fervent national pride and esteem, the architecture became a grouped emission of neo-nationalistic awareness unreservedly influencing the thriving and declining art of architecture and the methodology of dealing the tradition and modernity. Directed by the new regime, the "nationalistic style" became an implementation-must for architectural activities. Acculturated by nationalism and revivalism, the "traditional Chinese curved roof" stayed as a favorite. Of course, it now has a new mission to manifest the grand glamour of a socialistic nation. With vigor and vitality, the constructing guidelines of being "applicable, economic, beautiful, thrifty and industrious for the nation" was temporarily shelved. The situation did not change until the late 1950s. The sluggish in economy forced the architectural activities to adjust their course to be more practical by considering the economic factor, so be more abstinent and effective in designing the architecture with nationalistic, fashionable and regional characteristics by using contracted traditional or nationalistic symbols. This became a general practice in designing iconic or official buildings. From their formats, the decorous nationalistic Chinese official architecture style has resemblance with "Chinese modern architecture" in the republic era.

Fig. 3-53 Rendering of Beijing Planetarium

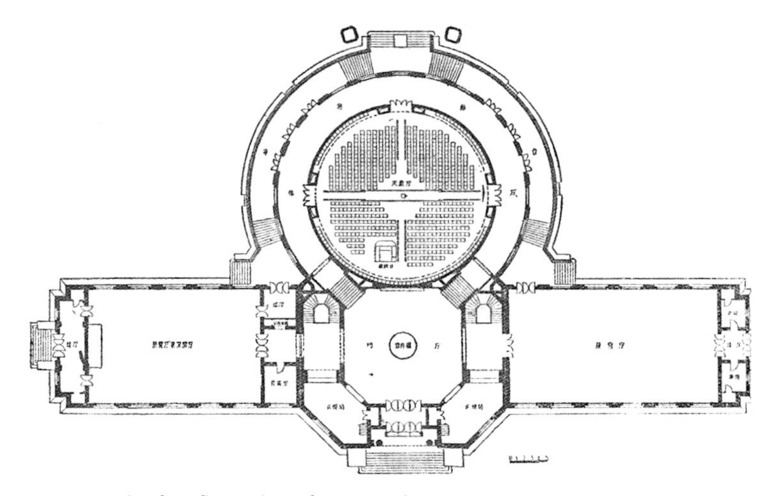

Fig. 3-54 The first floor plan of Beijing Planetarium

Case 1 Beijing Planetarium

Beijing Planetarium is the first building to demonstrate the astronomical phenomena. It was built in the same period as the well-known Beijing Ten Great Buildings. It was designed by architect Zhang Kaiji. The construction work was done during 1956-1957. The design was centered on the dome of the Celestial Theater, which is simple and elaborated with character (Fig. 3-53).

Occupying an area of 2.5 hectares with a construction area of 3,500 square meters, situated in the south bound of Xizhimen Street in Beijing, the planetarium, divided into Celestial Theater, auditorium, and exhibition hall, is to educate the populace with astronomical knowledge and with a demonstration of the celestial projection. The architectural layout was symmetrically arranged by functions with cascaded axes (Fig. 3-54). The octagonal lobby rises in the center of the front façade. It is a hub and an exhibition hall with a 10-meter high Foucault pendulum (an instrument demonstrates the rotation of the Earth). The two wings by the lobby are the exhibition hall and the auditorium. The lobby connects

Fig. 3-55 The dome of the Celestial Theater in Beijing Planetarium

directly to the main body—the domed Celestial Theater. The hallway surrounding the Celestial Theater connects to the traffic. The plan layout of the Celestial Theater is circular. The dome has inner and outer layers. The outer layer has a diameter of 25m, which is constructed with a thin shell structure in ferroconcrete. The inner layer has a diameter of 23m, which is in a hemispheroid simulating the celestial concave. The Theater has 254 seats, with a planetarium made in China at the center to display the sun, moon, constellations, meteors, comets, eclipse, lunar eclipse, and other celestial phenomena (Fig. 3-55).

The façade presentation is done by contents exhibited, where the lobby erects higher and it is flanked by balanced two wings with exterior walls covered by artificial granite. Right behind the Theater is the dome of the Celestial Theater. The overall layout is in classical western architecture style, sedate and

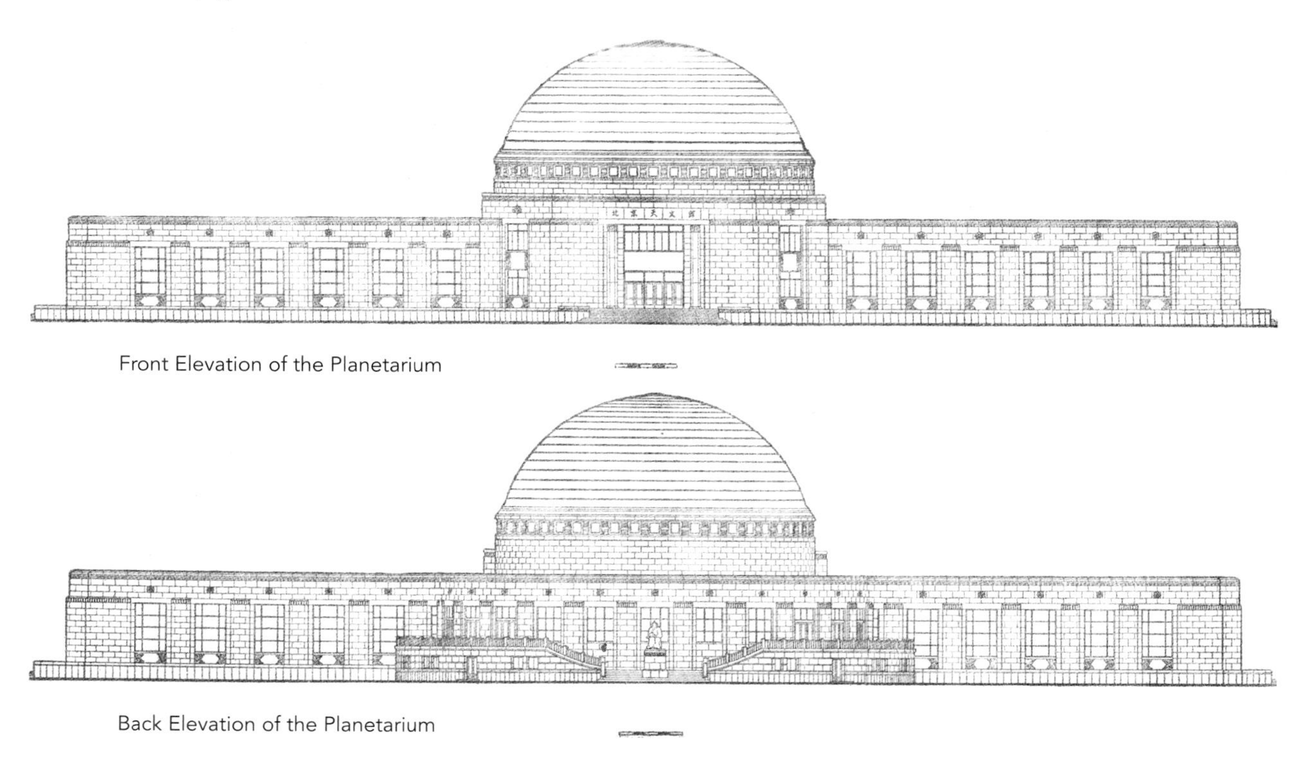

Fig. 3-56 Front and back elevations of Beijing Planetarium

Fig. 3-57 Partial front elevation of Beijing Planetarium

Fig. 3-59 Beijing Planetarium and an extension newly-built behind it

Fig. 3-58 Entrance to Beijing Planetarium

Fig. 3-60 Night scene of Beijing Planetarium

elegant. The wall and cornice have the traditional Chinese cloud patterns relating human and the celestial heaven (Fig. 3-56, Fig. 3-57). The architecture produces artistically appealing proportion and composition with accuracy, harmony and elegance, presenting the primordial, symmetrical and eternal universe described in classical astronomy with aesthetics that is valued by Classicism. The Chinese appreciate "heavenly fullness and earthly unassumingness". The planetarium is vaulted with a semi-spherical dome in coherence with not only the content it conveys but also the Chinese tradition.[1]

Furthermore, the architectural work integrates sculptures, paintings and architecture with organic integrity. The architrave of the front door has a white

Fig. 3-61 Celestial Theater of Beijing Planetarium seen from the rear extension

1 Zhang Kaiji et al. (1957), Beijing Planet*arium*, Architectural Journal (Issue 1)

marble continuous relief created by a famous artist (Fig. 3-58). The simulated zenith of the lobby has a 100 square meters gigantic color mural "Ancient Chinese Astronomical Myths" backgrounded with auspicious cloud patterns that it introduces well-known ancient Chinese myths of Chang-Er Fleeing to Moon, Hou Yi Shooting down Suns, Cowherd and Weaving Girl, Nüwa Repairing Wall of Heaven, Kua Fu Chasing Sun, and essences in ancient Chinese astronomy. The exterior and interior landscapes have many astronomical statues and meteor samples.

A new 20,000 square meters planetarium was built behind the old planetarium and open to the public in 2004. The old planetarium is shaped in heavenly vault and the new one, according to Einstein's principle of relativity that a massive celestial body will warp the time and space in continuity, therefore, made with a bold attempt to simulate the time-space distortion by warping its special hyperbolic glass curtain wall in a large arc, creates a background to compliment the old planetarium (Fig. 3-59 to Fig. 3-61).

Case 2 Beijing Minzu Hotel

In the forest of high-rises and flocking star-rated hotels in the west of Beijing Chang'an Avenue, not sumptuous and flashing, but busily and gracefully welcoming its guests, the building sitting nearby Fuxinmen has glorious history yet still strives for its vanguard renown. It is one of the "Beijing Ten Great Buildings" designated at the 10th anniversary of New China, Beijing Minzu Hotel, lasting over half a century and still radiating its glamour (Fig. 3-62).

Beijing Minzu Hotel in 48.4m high, with 12 stories and a construction area of 34,100 square meters was built in 1959. The design and construction project was presided over by the renowned architect Zhang Bo of the Beijing Institute of Architectural Design (BIAD). The ground level is for public facilities designed around the lobby and hallway. The Hotel has nine standard stories (Fig. 3-63). The building's plan is in "F" shape so that its main body is facing street and toward the south. The façade is designed with character and by functional requirements, technological availability

Fig. 3-62 Exterior of Beijing Minzu Hotel

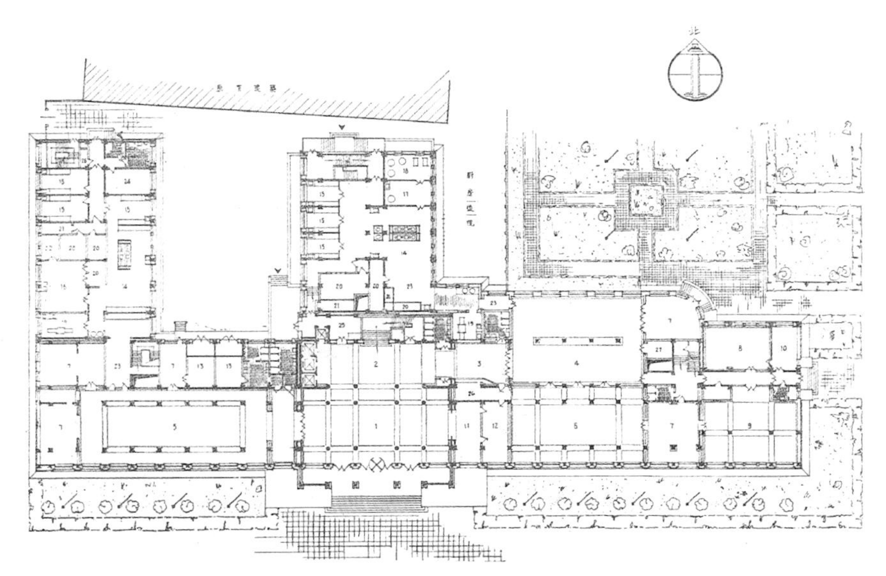

Fig. 3-63 The first floor plan of Beijing Minzu Hotel

and nationalistic traits. According to the basic characteristics of the frame structure, the entire façade is divided into ordered grids with the integration of story-layer dividers and exposed columns to animate its temperament and modernity. Columns and exterior walls are covered with light yellow brick tiles interlaced horizontally and vertically (renovated to have stone tiles). The exterior of the ground level is covered in granite with part of it in a bush-hammered finish. Balconies and terrace on the second level are designed with simple oversail eaves and Chinese balustrades. The ground level granite finish has the upper section lined with wide sash and the lower section added with base (Fig. 3-64 to Fig. 3-66). The entire façade is simple, lucid, and unpretentious and toned with nationalistic taste. The entrance vestibule has eight pieces of trelliswork, resembling the "lattice window" of ancient Chinese garden verandas, added with new contents, done by contemporary artists, exhibiting China's booming industry, agriculture,

Fig. 3-64 Entrance to Beijing Minzu Hotel

Fig. 3-65 Details of the exterior of Beijing Minzu Hotel

Fig. 3-66 A corner of Beijing Minzu Hotel

Fig. 3-67 Details of lattice decorations at the entrance to Beijing Minzu Hotel

transportation, science and technology (Fig. 3-67). The building integrates modern functionality, technology and the traditional Chinese culture, not by crowning but by unpretentiously and specifically implementation in a more flexible and cost-effective way.

In "Beijing Ten Great Buildings" celebrating the 10th National Day, the Minzu Hotel had its construction project commenced the last. After the rest projects entered the decoration stage, the Hotel just finished its foundation. With such a tight schedule, it was not only completed on time but also made several firsts in the architectural history of New China. It was the first building using pre-fabricated ferroconcrete frames for a 12 stories high building; the first construction project using only 120 working days from the rigging to finish installation; the first civil engineering example breaking the construction speed record; and, the first tallest building using industrialized construction method in China. In September 1959, Beijing Minzu Hotel was finished ahead of the schedule and commissioned into operation. It is eye-catching on the capacious Chang'an Avenue and one of the landmarks of Beijing City. It also witnessed the historical event of the normalization of relations between China and the United States.

Fig. 3-68 Exterior of the Capital Theatre

Case 3 The Capital Theatre

In Beijing's cultural life, the Capital Theatre is a noticeable landmark. In over 50 years span, it witnessed China's rise and carries unforgettable memories of several generations (Fig. 3-68).

Situated at Beijing Wangfujing Street, the Capital Theatre was the first theater built for professional drama play after the founding of New China. It also can be used to host large musical settings, theatrical dramas and movie plays. It was constructed in 1955 by the US-trained well-known architect Lin Leyi. The Theater occupies an area of 0.75 hectares of which the construction area is 11,500 square meters. The plan layout focuses its weight around the central axis and has major functional spaces, including lobby, auditorium, performing stage and rehearsal studio, aligned with it. The auditorium is rectangular with 26m in length and 12.5 in height. It can accommodate 1,302 seats all with clear perspective. The ceiling is decorated with large-size festoon at the center (Fig. 3-69, Fig. 3-70). The stage is 20m in depth with well functional front stage and backstage that are convenient to use, especially, the revolving stage of 16m in diameter, which is the first of its kind designed and constructed by Chinese. The Theater has a spacious refresh hall decorated to produce classical atmosphere, showing its

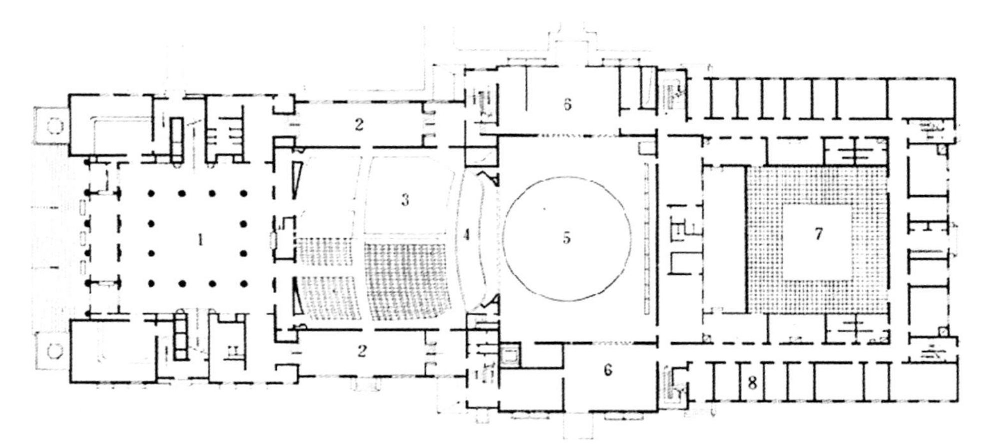

1. Lobby 2. Lounge 3. Auditorium 4. Orchestra Pit
5. Stage 6. Stage Side 7. Drill Hall 8.Dressing Room

Plan of First Floor

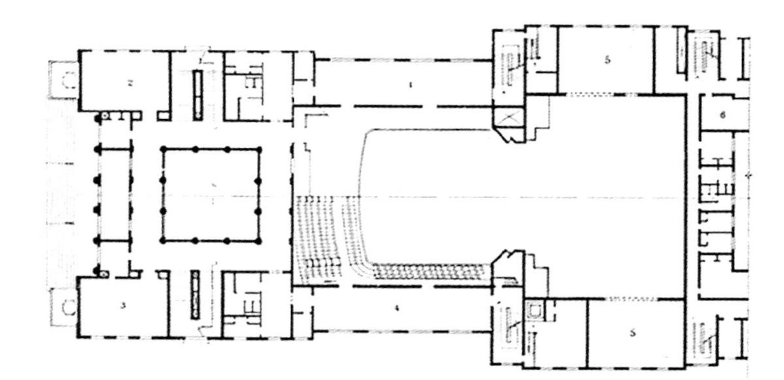

Plan of Second Floor

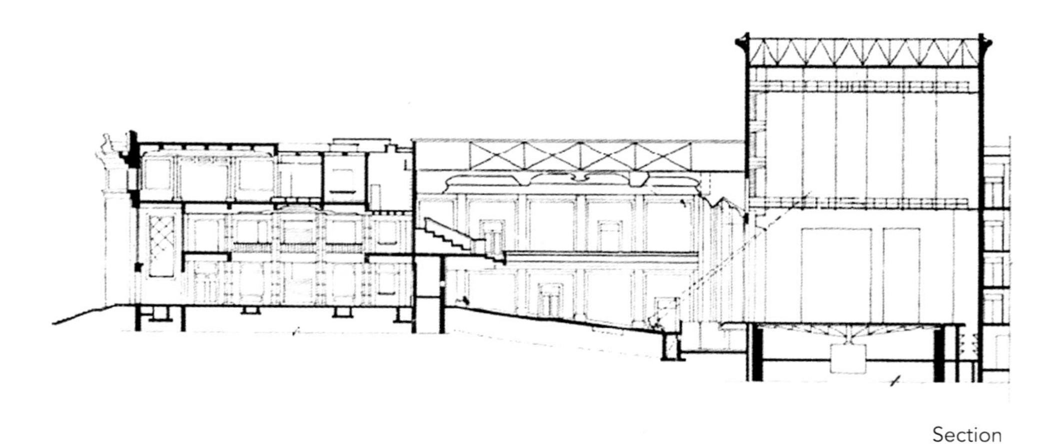

Section

Fig. 3-69 Ground plan and profile of the Capital Theatre

Fig. 3-70 The ceiling of the auditorium in the Capital Theatre

Fig. 3-71 Interior of the Capital Theatre lobby

Fig. 3-72 Eaves of the front of the Capital Theatre

grandeur and magnificence and lightened by various lighting fixtures to deliver a peaceful elegance.

When the Theater was in its construction, it was the time that China took the "unjustifiable bias" stance toward the former Soviet Union. The architectural design stayed in the same course. The plan and external appearance of the Theater resemble the Alisher Navoi Opera and Ballet Theatre of Central Asia. The main composition is demure and graceful similar to Russian architectural style. The exterior

Fig. 3-73 Detailed pattern decorations on the elevation faces of the Capital Theatre

Fig. 3-74 Entrance to the first floor of the Capital Theatre

walls are built with bricks, mortar, artificial stones and moldings to simulate the heaviness of a stone building. However, it integrates the characteristics of the details of the traditional Chinese architecture symbols, such as ornamental column, screen wall, *queti*, architrave, coffer and embossed painting, which helps the modern architecture emitting heavy nationalistic glamour (Fig. 3-71 to Fig. 3-74). The building demonstrates the designer's effort in breaking the revivalism by not using the traditional Chinese curved roof, upturn eaves and bright-red columns but using well functioning, elaborated details and exquisite decors. It is one of the exemplary theatrical buildings of New China.

The Capital Theater was turned over to Beijing People's Art Theatre in 1955. After that, a series of popular dramas was played there, such as *Teahouse*, *Thunderstorm* and *The Best Restaurant in the World*. It has been playing an important role in the development and flourishing of the art of Chinese drama. Its architecture received the outstanding creation award from the architectural society of China for its extraordinary ingenuity.

Fig. 3-75 The Administration Building of the Ministry of Construction and Engineering of the People's Republic of China

Fig. 3-76 Entrance to the Administration Building of the Ministry of Construction and Engineering of the People's Republic of China

Case 4 Administration Building of Ministry of Construction and Engineering of the People's Republic of China

In the late 1950s, Chinese architecture often gesticulated in eclecticism by using modern architectural structure and function as the basis complemented with decors and patterns to exhibit the nationalistic style. The Administration Building of Ministry of Construction and Engineering (today's Ministry of Housing and Urban-Rural Development of the People's Republic of China) is one of the exemplary works (Fig. 3-75).

The building is at Baiwanzhuang in the western suburb of Beijing. It was designed and presided over by the well-known architect Gong Deshun of Beijing Industrial Design and Research Institute, the Ministry of Construction and Engineering. It was constructed during 1955—1957. It is one of the large-scale ministry office buildings of New China. The building occupies 10 hectares of which 38,000 square meters were used as the construction area. The building has one underground level and seven levels

Fig. 3-77 Details of the entrance to the Administration Building of the Ministry of Construction and Engineering of the People's Republic of China

Fig. 3-78 External wall of the Administration Building of the Ministry of Construction and Engineering of the People's Republic of China

above the ground. The building is in brick-concrete. The height of the building earned its reputation of a technological innovation. Its design integrates the functional requirements with structure criteria. It uses the flat roof. The exterior walls are in granitic plaster finish. The dealing of the mass and elevation façade are clear. The main façade is in composition of neo-classicism, which is grand and towering. The elevation is divided into three sections. The azimuth is symmetrical. The lobby has stone structure mimicking beams, architraves, roof and decors of the traditional Chinese wooden structure. A high stairway leading to the front entrance generates an effect of balanced yet majestic and antiquing simplicity and sophistication (Fig. 3-76). The front perspectives, viewing from the south toward the north, the 38super large windows look grandly magnificent and sedately solemn. The eave mouths in the borrowed style of stone oversail eaves from the traditional Chinese style are projected by using ferroconcrete rafters balanced beautifully with excellent details. Furthermore, the interior uses contracted styling of the traditional Chinese architectural elements and abstracted patterns, maintaining part of the nationalistic style yet revealing the trace of a successful transition in the late stage of it[1] (Fig. 3-77, Fig. 3-78). In the 1980s, the building was recommended by the Royal Institute of British Architects (RIBA) to be one of the 43 global outstanding office buildings of national institutes. Up to the time of writing, it has been in use over 50 years and historical waning and waxing to have not slashed a bit of splendor of its grace and elegance.

1 Zou Denong (2010), *Chinese Modern Archiecture* (pp 52), Beijing: China Architecture & Building Press

Chapter 4
Folk Wisdom

Folk Culture's Take on Architectural Integration of Chinese and Western

4.1 Exemplar of Integration of Chinese and Western Culture: Lilong Dwellings

4.2 Featured Private Residences

4.3 "Western-Style Façade" and "China's Baroque Architecture"

There were two takes in the integration of architectural culture of Chinese and Western in the modern history of China. One was to integrate the traditional Chinese architecture with western architectural technology, patterns and modeling characteristics; another was to integrate the western architecture with the traditional Chinese architectural modeling characteristics, decors and patterns. From culturology, both takes demonstrated the profound rationality of the Chinese traditional ideology that architecture is a tool to create living quarters for the people and establish social order.[1] The former is a practice to build dwellings and shops by non-professionals who often lacking architect's profundity yet with wisdom of daily life and the latter is participated by professional architects who create buildings of administrative, financial, school and public edifices with eclectic touch or neo-nationalistic persuasion.

Although they are architectural products of the same acculturation of Chinese and Western, the one formed from the natural instillation is different from the articulated neo-nationalistic or eclectic architecture by architects. The "dynamic sensation of Chinese art and Western science", as their ingenuity basis, created by non-professionals was still cored on the traditional Chinese architecture and only built with western architectural technology, patterns and modeling characteristics. In a sense, the take of insisting on Chinese architectural culture as the core was subliminal and unresisting, while, the imbibing and embracing the western architectural culture were assertive and welcoming. This take, in China's overall modernization process, was gaining momentum in popularity. Most of all, it was well accepted by the populace across the mainland, which was evidenced by the number of private dwellings in both urban and rural areas, as milieus in Chinese modern architecture history. Fore example, in the second half of the 19th century, the swarming population in Shanghai concession sparked off the brick-and-wood structure genre of Shanghai Shikumen Lilong dwellings around 1870 to meet the housing demand of newcomers, which, in fact, was a variant of the traditional Chinese Sanheyuan (attached three-compound surrounding an open courtyard) turned into a new type of urban dwelling style. Although, not designed by an architect, the genre is different from Sanheyuan in its two-story structure with attached compound layout. Phasing with time, more of them were built with the western architectural materials, structure, construction techniques and decorations. They only retained the portions for ritual discipline and ethical order in the traditional Chinese house plan layout—assertively adopting materials, techniques and decorative elements of the western architecture and inheriting the space design of the traditional Chinese architecture. Similar examples can be found from Kaiping Watchtowers in Guangdong hometown of overseas Chinese in the south, shop-dwellings in Harbin Daowai district, "China's Baroque" architecture in the north, and manors of rich merchants across the urban and rural areas of the country.

Chinese have long been inclining to use architecture as a social order formation effectuated by establishing the architectural elements with the user's innate traits, such as orientation, location and interrelationship with the nature, where the architectural elements are deterministic factors. Therefore, within the traditional Chinese framework, the architectural plan layout is the focus, which is presented evidently in Shanghai Shikumen Lilong dwellings and Kaiping Watchtowers, as an adaption to the occidental civilization. The assertion on retaining the traditional Chinese architectural layout is the core essence of Chinese architectural ideology and culture.

The insistence on the core essence of architecture makes the Chinese seemed conservative and

1 Wang Lumin (2006), *Distant Perception—Study on Two Approaches in Modern Integration of Chinese and Western Architecture*, New Architecture (Issue 5 pp 54)

uncompromising, yet, when incorporating the western architectural elements, the Chinese are natural and capricious. The work would be designed at liberty by the craftsman according to the sole preferences of the owner to stay in with the general aesthetic tastes and fashion with honest and straightforward presentation. Although non-professional, tawdry and pretentious, it would be accepted by the populace. Greece Romanesque, Gothic, Renaissance and Baroque styles were stewed into a "potpourri", especially the Baroque style that was prominent and often doped with the traditional Chinese architectural decorative details to become a baroque variant - "China's Baroque" style. The differences between traditional and modern buildings in a large degree are in structures, materials and construction methods, where Baroque style happens to be flexible with high inclusivity. It can be done with brick-and-wood, bricks, or modern ferroconcrete by Chinese craftsmanship. In fact, the Baroque, when imported from Italy to other countries in Europe and America, it was recreated into variants also, such as that in Germany, Austria, France and Spain. Therefore, it was not surprising to see a variant of Baroque in China. In addition, there is a concordance between the traditional Chinese architecture and the Baroque. The former learned early to use upward curving eaves and corners to deliver dexterity and the latter emphasizes creating kinetic dynamics. The former's rich and diversified detail designs and the latter's over-elaborated carvings share commonality. This makes the two a natural combination. Tablets, *quetis*, balusters, festoons, hanging lotus and cloud patterns are easily to be incorporated into Baroque's architecture elements. Therefore, the Baroque reconciled several paradoxes encountered during the process of Chinese modern architecture transition and effortlessly, gained wide acceptance by the populace.

In general, private buildings in modern China inclined to integrate the western architectural elements into the traditional Chinese architectural composition. The absorption of the western architectural elements was driven by fashion that was a reaction of the society to the invading foreign culture, which then was recreated with folk wisdom to become a unique style unveiling the transition evidence of Chinese modern society. The wisdom also left us with valuable folk architectural culture heritage.

4.1 Exemplar of Integration of Chinese and Western Culture: Lilong Dwellings

The main thread in China's modern architectural experience was the aspiration of the nationalistic style. In pursuing such, besides efforts of professional architects and scholars, large amount of works in the architectural integration of Chinese and Western were echoed through the propagation channel of the populace. In some aspects, this echo happened much earlier with a wider rippling effect. Especially, to the urban dwellings and commercial buildings, Lilong dwellings became an exemplar to the general public in Shanghai, Wuhan, Hangzhou, Tianjin and Qingdao, harboring generations' sentiments of the homeland, while, as an architectural phenomenon, they were studied by the academia. Lilong dwellings were a product of acculturation and technological exchange. They represented a dwelling genre developed after China shifted into semi-colonial and semi-feudal society that had its low-level attached compound influenced by the western modern architecture. It was one of important architectural genres in Chinese architectural or residential history, which initially began in Shanghai in the late 19th century. Before the founding of New China in 1949, it already became the most popular dwelling style in Shanghai, Tianjin and Wuhan.

Case 1 Shanghai Shikumen

Shikumen is the modern residential architecture with the most Shanghai features. As an exemplary architecture of the time, to Shanghai Bund, Shikumen cluster of the Old Shanghai unfurled Shanghai's

Fig. 4-1 Bird's eye view of central Shanghai in the 1930s

heterogeneous cultural characteristics. It was created in the early 70s of the 19th century. The Taiping Rebellion (or the Taiping Heavenly Kingdom) uprising forced the wealthy gentries in Jiangsu and Zhejiang provinces to move to Shanghai concession to seek for shelters. Foreign realtors seized the business opportunity to build a large number of dwellings. Shy on available urban land lots, those dwellings were constructed as attached compounds (Fig. 4-1, Fig. 4-2). After entering the 20th century, the genre of the architectural integration of Chinese and Western of Shikumen became the mainstream during Shanghai development rush. The building used stone materials as the door frame, black-varnished thick solid board as door leaves added with a pair of door rappers that gained its name "Shikumen" (Stone Warehouse Door).

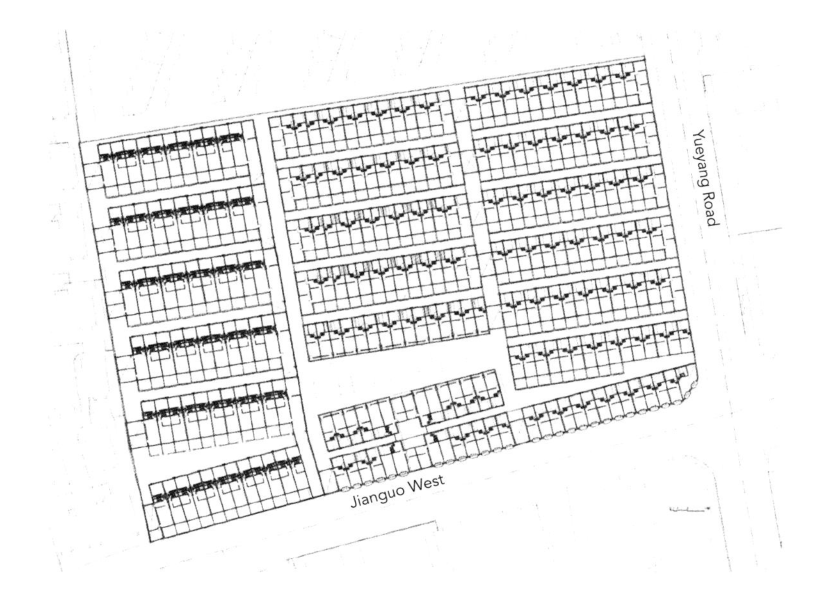

Fig. 4-2 Master plan of Jianyeli, Shanghai

To satisfy the traditional Chinese residential style, Shikumen, except mimicking partially the attached compound, has the rest layouts in the traditional Jiangnan private house style, which is usually in sizes of three-bay or five-bay, retaining symmetry along the central axis. A rectangular front courtyard with its longer side in parallel and right behind the entrance is followed by the living room and right after that is a rear courtyard with a depth only half the size of the front courtyard, where there is a well. Right behind the rear courtyard is an amenity with a single layer sloping roof, used as a kitchen, multi-purpose spare room or

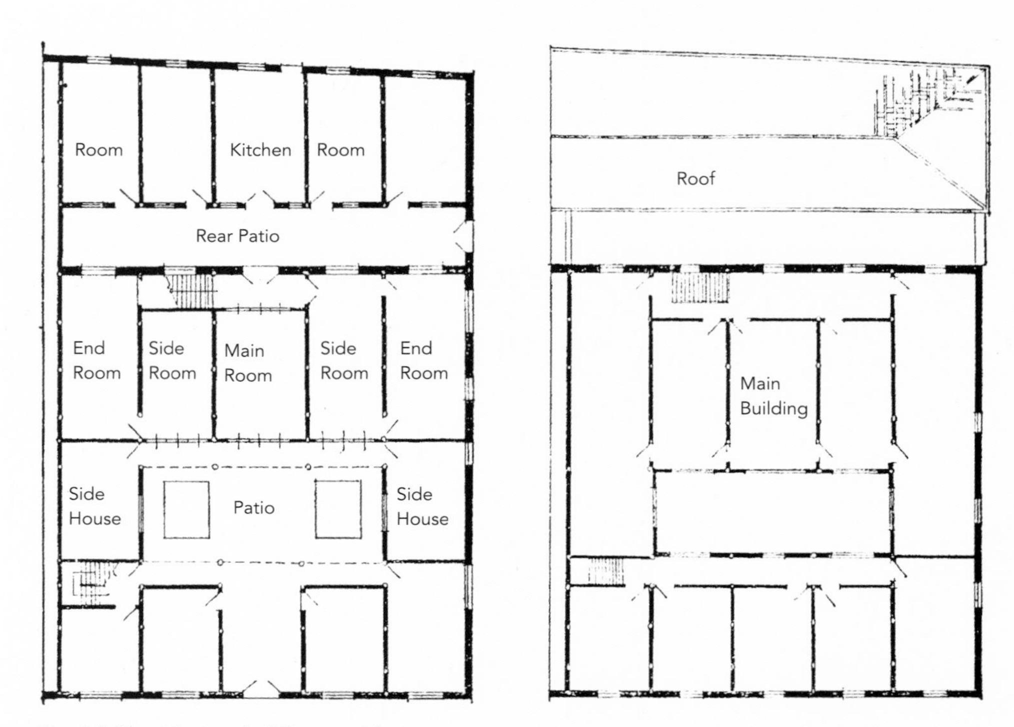

Fig. 4-3 Plan of an early Llilong residence

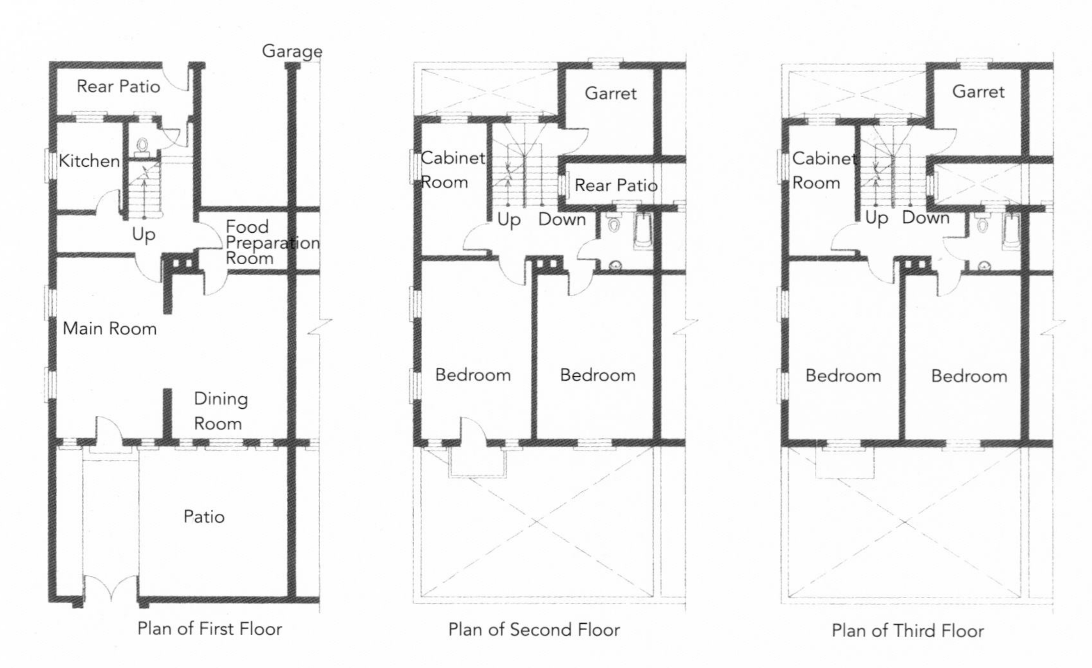

Fig. 4-4 Plan of a late Llilong residence

Fig. 4-5 A Llilong primary school in a drawing

Fig. 4-6 A road in Bugaoli, Shanghai

storage room. The two sides of the front courtyard and living room are for the left and right wing rooms. The second floor has the same layout as the ground level, only the kitchen has a loft "garret" and a flat roof on its top (Fig. 4-3). The entire dwelling has entrances at its front and back. The front façade is the combination of the courtyard wall and the gable wall of the wing room, with "Shikumen" at the center. Although the dwelling cluster is usually surrounded by busy streets, it is unruffled from the noisy city by the wall and courtyards. The genre was likened by the then wealthy gentries.

From the early of the 20th century and onward, Shanghai families downsized, so did the lifestyle and Shikumen layouts. The new Shikumen architecture still retained its space arrangement and enclosing walls but not the carvings, rather, being concise and most of them were "one-bay" in size (no wing room), which was suitable for small sized families. Some were "two-bay' in size (one living room and one wing room) to meet the needs of families with more members (Fig. 4-4). The alley was in 4m width. Most of the dwellings were in 2—3 stories. A loft or garret was built on the stair landing. The front façade had a balcony. Those ones built after 1920s all had modern sanitary fittings. After the 1930s, because of Shanghai housing shortage, some residence let out their extra rooms that changed the original purpose of the design to accommodate more than one families in one dwelling, creating an enclosed neighborhood under the same roof (Fig. 4-5 to Fig. 4-7).

Fig. 4-7 Shikumen Lilong, Shanghai

Fig. 4-8 Entrance I to a Lilong in Shanghai

Fig. 4-9 Entrance II to a Lilong in Shanghai

Fig. 4-10 Exterior I of a Shikumen residence in Shanghai

Fig. 4-11 Exterior II of a Shikumen dwellings in Shanghai

Shikumen architecture that incorporated much of the western architecture style is rich in decorations. The dwellings share commonality and behavior. Most of the exterior walls are brick walls without plastering built with blue or red bricks, and some are mixed with limed join. Gable walls and dwellings at the alley entrance are heavily decorated. The early Shikumens often had fire-sealing gable (or Corbie gable) or *guanyindou* and later changed to use the western styled bargeboard (Fig. 4-8 to Fig. 4-11). The decors on the architrave of early Shikumens were similar to those carvings on blue brick tiles on the door canopy of Jiangnan style and some were in auspicious characters. Door frames were boulder strips. On the sides of the architrave were festooned *quetis*. Later, Shikumens were influenced more by the western style began to have imitated western classical architecture decors of triangles, semicircles, arcs or rectangular floral designs that they were in elaborated and diversified varieties. These are the most featured group built in Shikumen style (Fig. 4-12).

Shikumen is the most popular, massive in numbers common dwellings in Old Shanghai. Even in today, there are still a large number of them remained. Shikumen with the courtyard layout of Jiangnan style and the European attached compound structure is the result of the architectural integration of Chinese and Western. It was further expanded not in construction size but in household-wise to become a shared dwelling of families, divided but not separated, forming a small neighborhood under the same roof, sedated from the outside but not the inside. As one of the amazing architectural genres in China's modern architectural history, it creates its own residential culture, which opens another chapter for the regional architectural culture of Shanghai and has its traits recorded.

Case 2 Wuhan Lifen

Lilong dwelling is called Lifen in Wuhan. It is a synonym of Shanghai's Shikumen. It was developed by Shanghai realtors. Therefore, its style is very similar to that of Shanghai's Shikumen. Lifen dwelling is an

Fig. 4-12 Diverse architraves of Shikumen dwellings in Shanghai

epitome of near century-old residential culture of Wuhan. It is also an integration of the western attached compound style and the traditional Chinese courtyard architecture after Hankou became an open port. It is also an important medium unfolding the unique regional architectural culture of Wuhan (Fig. 4-13).

Most of Lifen dwellings were constructed during the late 19th century and early 20th century before 1937's Anti-Japanese War. In over 30 years, 208 alleys of Lifen dwellings were built in total, mostly in Hankou. They can be categorized in two types. Dwellings built by locals were in wood structure with less space, poor planning, less domestic amenities and less sophistication. Another type was those built jointly by rich merchants resembling early real estate development project with sound planning. They were built in two-level brick-and-wood attached compound structure, uniformly styled. The center was the living room, wing rooms were on two sides, stairs and kitchen were in the rear and some had a backyard. The hallway was spacious. The dwelling was well equipped with domestic facilities, such as those in Tongxinli, Hanrunli, Shanghaicun (Fig. 4-14 to Fig. 4-18). After becoming a trade port and concessions in 1861, Hankou experienced first the imported occidental civilization. Another striking foreign style Lifen emerged. This kind of Lifen had a higher two stories in mixed red and black bricks built in typical European style. Lifen dwellings were usually located in busy city districts and architected with serenity amid buzzing surroundings, a comfortable size and a friendly neighborhood settings.

Wuhan's Lifen conveys the architectural humanization consideration. In dealing with the "scorching city", Lifen dwellings were added with ventilation and sunshade fittings to moderate extreme seasonal conditions and alleviate summer indoor sizzling heat. Windows were made in two layers, an internal glass pane layer and an external wooden shutter. Screens joined by using wooden panels bearing

Fig. 4-13 Bird's eye view of Hanrunli, Wuhan City (From Li Baihao, Southeast University)

① Lane
② Entrance of Shengli Street
③ Entrance of Dongting Street
④ Main Lane

Fig. 4-14 Plan of Tongxingli, Hankou, Wuhan and the elevation of the current main lane (from Li Baihao, Southeast University)

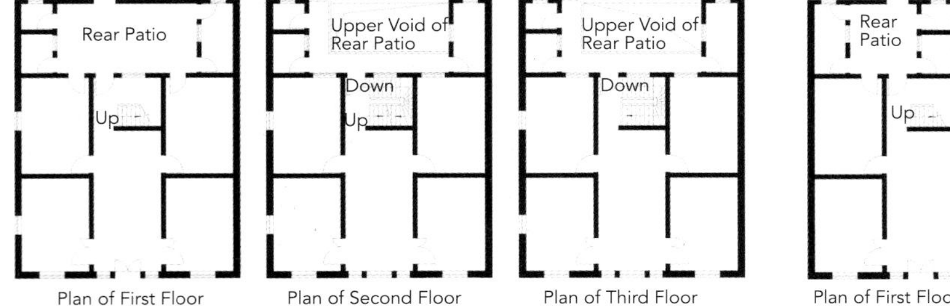

Fig. 4-15 Three-bay plan of Shanghai Village, Wuhan City (from Li Baihao, Southeast University)

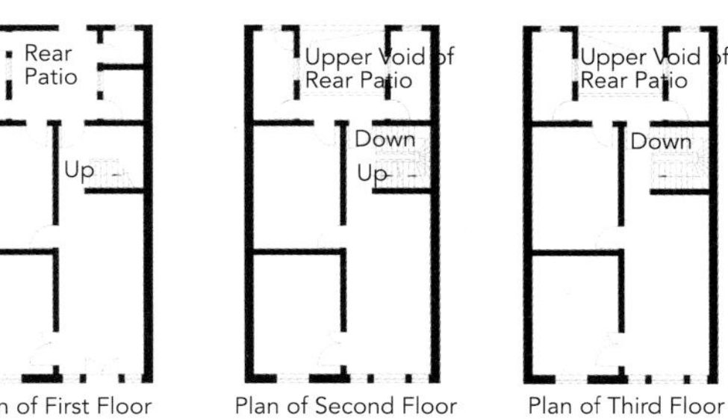

Fig. 4-16 Two-bay plan of Shanghai Village, Wuhan City (from Li Baihao, Southeast University)

Fig. 4-17 Main lane of Shanghai Village, Wuhan City (from Li Baihao, Southeast University)

Fig. 4-18 Second lane of Shanghai Village, Wuhan City (from Li Baihao, Southeast University)

Fig. 4-19 Door frame I decorated with geometric patterns (from Li Baihao, Southeast University)

Fig. 4-20 Door frame II decorated with geometric patterns (from Li Baihao, Southeast University)

Fig. 4-21 Plant-themed decorations on a lintel (from Li Baihao, Southeast University)

Fig. 4-22 No.9 window frame of Dongting Village, Wuhan City

Fig. 4-23 No.23 window frame of Dongting Village, Wuhan City

decorative carvings with elegance and primitive simplicity were used to separate the interiors. In winter, they were used as decorations and in summer, they were removed to return the entire interiors for venting in daytime and cooling at night.

Wuhan, diversified in culture, is inhabited by "people from all walks of life and ethnicities". Wandering through these outstanding Lifen dwellings, elements, inclusively, of Chinese memorial arch, western architraves or integration of Chinese and Western decorative designs (Fig. 4-19 to Fig. 4-21) are not only integrated with the western modern architecture but also recounted with new annotations of the architectural language, unfurling the creator's dexterity and ingenuity. The interior decorations are exquisite. Wooden trellis, balustrades, panels are all elaborated works. The patterns are mostly in the traditional ingenious and elegant Chinese style (Fig. 4-22, Fig. 4-23). These have created an architectural genre with unique cultural connotation as well as a historical memory.

4.2 Featured Private Residences

Compared with Lifen dwellings in other open trade ports and concessions influenced under the occidental civilization, the rest of urban cities, towns, rural villages were slow in modernization. Under the strong traditional cultural ideology, the westernization practice in rural regions, when competing with the architectural integration of Chinese and Western with Chinese culture emphasis, appeared to be insignificant that the best can be said is merely an exercise of the "dynamic sensation of Chinese art and Western science". Owners of private dwellings were most of official gentries of hardliners with deeply rooted traditional Chinese cultural ideology. Some of them, either out of curiosity or from overseas experience as an entrepreneur acknowledging the advance of the occidental civilization, chose to have a western living style for a psychological satisfaction or to match their status, rank or career convenience that was echoed in the building they endorsed. However, even the extreme capitalists and dignitaries admiring the western culture, they had their reservations in using the western architectural language. Regardless of pavilion, terrace, veranda, belvedere of landscaping architecture or private mansion, the overall design of bay modeling, room structure, lintel, tablet and couplet still conveyed the owners' true desire for the orthodox Chinese cultural heritage. The western architectural components were decorated with Chinese design patterns, revealing a strong contrast of the difference in aesthetic nature. This reflected exactly the contemporary Chinese social structure and cultural dynamics at that time.

Case 1 Kaiping Watchtowers

In the open field of Kaiping City of Guangdong, small castle-like buildings of the architectural integration of Chinese and Western scattering among the traditional peasant dwellings of rural villages in South China create an unusual rural landscape (Fig. 4-24). Kaiping Watchtowers are a special genre of China's vernacular architecture. The genre roughly started in the late Ming Dynasty (the 16th century). As a multi-story tower-type architecture, it had defense and residential purposes with integrated Chinese and

Fig. 4-24 Kaiping watchtowers in village

Western styles. Around the 1920s—30s, the development of Kaiping Watchtowers reaching its peak had over 3,000 in numbers with 1,800 extant today. Kaiping Watchtowers in its variety were large in scale and diverse in styles. They had integrated almost all appealing elements of various architecture genres extant at that time. In 2007, they were designated as the World Heritage Sites. Their critics recorded said, "an exuberant integration of architecture and decorations of Chinese and Western", "...revealing the significant influence of Kaiping overseas emigrants in Southeast Asia, Oceania and North America in the 19th century and early 20th century and the close bond between them and their family clans".

Fig. 4-25 Castle-like Kaiping watchtowers

Kaiping region is the well-known hometown of overseas Chinese. Bandits saw the returning overseas Chinese who were acquiring land properties, building their homes and raising a family as their "preys". "Wealthy families were forced to use iron rods, rocks and cement to build three or four–story towers to defend themselves, while, ones with less financial surplus gathered together in one tower". Therefore, Kaiping Watchtowers erected as self-defense fortresses and dwellings for the returning overseas Chinese. This type of building in layout and structure was designed for that purpose. The exterior walls were built to be thick and strong with bricks, rocks or concrete. The front door used heavy steel board. Windows were enforced with metal grilles. The top level was built with overhangings to serve as a watch tower. Embrasures were put on four sidewalls. The design was proven to be effective historically in defending lives and properties of the villagers (Fig. 4-25).

Fig. 4-26 Kaiping watchtowers I

There are three types of Kaiping Watchtowers. Watch tower or light tower (or *genglou* or *denglou*) built around the two gates of a village or on the top of a hill were used by the territorial militia and the watchmen, with stock of weapons, searchlight and alarming apparatus to warn villagers of approaching hostiles. Communal Tower built by a dozen or less families, in three—six-story high and with 2—4 rooms per story were used as the emergency shelter for the owning families when there was foray or flood. The

Fig. 4-27 Kaiping watchtowers II

Fig. 4-28 Kaiping watchtowers III

Fig. 4-29 Kaiping watchtowers IV

third type was the residential tower built by a single returning overseas Chinese family as a residence. The external appearance of Kaiping Watchtower is the architectural integration of Chinese and Western with variety of design, including winding corridors, terrace, castle-like structure, and mixed (Fig. 4-26 to Fig. 4-29). The most amazing characteristic is the integration of different foreign architectural styles in one tower. Villagers brought medieval castle design to the country side of Lingnan and integrated them with Doric colonnades and Gothic arch, or even Baroque's flamboyant laurel with exquisite decorations of Corinthian order acanthus and graceful Ionic order volute. Window casing, lintels and pediments all reveal great originality. Some towers even have the traditional Chinese plastered reliefs and oversail eaves. These buildings are ambitious and extravagant to include every decorative essence from the ancient Greece to European Renaissance of the occidental civilization and from the ancient China, India and Islam of the East into one. These elements integrated harmoniously are still effusing unique artistic glamour and magnificence after centuries of vicissitudes (Fig. 4-30 to Fig. 4-32). Kaiping Watchtowers although with the western castle look and decors have their plan layout in the three-bay local tradition. The lineage can be traced back to Lingnan tradition of two-bay Mingciwu (living room-room-amenity) having the front facing the south, which has the custom of placing the ancestral memorial tablet at the side facing the front door in the living room at the lowest level or the top level. Up to now, no design drawings have been found for Kaiping Watchtowers or proofs of trained architect involved in designing the towers. From the non-professional construction craftsmanship for the western architectural elements, it could be assumed that the work was a creation from a combination of images from pictures or paintings from overseas, the owner's ideas, the craftsman's own imagination and construction experience.

Dexterously, Kaiping Watchtowers reflect the integration of the traditional Chinese vernacular architecture and the western architectural culture, unfolding the acceptance of the integration Chinese and Western cultures in modern time, which have created a new landscape for the architectural art of the world. Actually, the real value of Kaiping Watchtowers is the implication of its architecture. Ostensibly, it is

Fig. 4-30 Western decoration on Kaiping watchtowers

Fig. 4-31 Time-honored Kaiping watchtowers

Fig. 4-32 A corner of the external wall of a Kaiping watchtower

a manifestation of the culture of returning overseas Chinese. From a broader perspective, it is an epitome of the world immigrant culture resulted from transnational and cross-regional migration.

Fig. 4-33 High wall of the waterside Zhang Shiming's Former Residence

Fig. 4-34 Entrance to Zhang Shiming's Former Residence

Case 2 Zhang Shiming's Former Residence, in Nanxun Town, Huzhou, Zhejiang Province (Yide Mansion)

Yide Mansion, the former residence of Zhang Shiming[1] in Nanxun Town, Huzhou, Zhejiang Province, is a typical manor of Jiangnan's peerages. Inside the manor property, thousands stone sculptures, brick carvings, wooden carvings, glass carvings are categorized as "The Four Ultimates", among them, the "Western Ballroom", Plantain Hall are mostly known. It is surrounded by a creek on three sides. Its deeply layered inward courtyards have modest appearance yet gorgeous landscape inside. The high entrance gate tower is unassuming. The manor sitting on a quiet riverside appears as an unvarying rustic whitewashed wall topped with dark tiles. Its reflection in the creek is as a black and white photo (Fig. 4-33, Fig. 4-34). All the "fortune haps" and architectural pith are wrapped inside the walls. The building has the mixed Ming and Qing styles as its main note, the European style as the minor. It is a classic of the architectural integration of Chinese and Western, flashing high artistic, architectural and cultural value.

The manor facing the east occupies a land of eight *mu* (about $5,333m^2$). The front is Guxun Creek. The construction area is 7,000 square meters and has five-rows of courtyard in the longitudinal direction. Along its central axis, side yards and cross yards are irregularly arranged. There are still 244 rooms survived after a century. No matter in scale or luxury, this manor belongs to one of the rare gigantic real

1 Zhang Shiming (1871—1927), given name Junheng, a native of Nanxun in Huzhou of Zhejiang, was an imperial designated provincial scholar of the 20th year 0f Guangxu reign of Qing (1894). He was the eldest grandson of Zhang Songxian, one of the wealthies, and the elder cousin of Zhang Jingjiang, one of KMT senior statemen. He was a known collector of ancient scrolls, tablet inscriptions and rare rocks, one of the four book collectors of Nanxun in the late of Qing and the early Republic era. He was also the founder and sponsor of Hangzhou Xiling Printing House, a good friend of Wu Changshuo and Mao Fu'an, known scholars. Besides being a collector, he built the Yide Mansion, known for its grandiosity, delicate carvings and the integrated architecture of Chinese and Western. He was kidnapped and blackmailed in 1925, and died in 1927 aftersuffering of the aftermath of the atrocity.

Fig. 4-35 Main hall of Zhang Shiming's Former Residence —Yide Mansion

estates of Jiangnan's glorious families. Its appearance has not changed much. The magnificent style, grand structure, elaborated craftsmanship and sophisticated architecture deserve the reputation of the "Best Jiangnan Dwelling".

Zhang's old residence has profound courtyards, winding verandas and a rolling skyline. The front courtyard is an Erheyuan (one main with one attached wing room), the second and the third courtyards are Sanheyuans. The Erheyuan with a palanquin porch has four bays. A brick Ruyi gate tower with an enchased stereoscopic "Deities Celebrating the Birthday of Queen Mother of the West" connected to the porch. Right behind the gate tower is the main hall "Yide Hall", towering and expansive. A multiple folding door is placed in the back of the hall, which can be opened to give more space (Fig. 4-35). The Second courtyard of one-hall and two-wing-room congregation is the "Ladies' Court", used for reception, work and rest. Its decorations are recherché. Glass windows on the second floor are put up on the side to face the courtyard and inlaid with a set of appealing French hand-cut blue-crystal glass in rhombus shape with flowers and fruits (Fig. 4-36, Fig. 4-37). The so-called Plantain Hall is actually a courtyard with balustrade, door frames, and lattice window carved with exquisite designs with large and curly leafs in western style, properly preened in dark emerald. It was once said that circular dew drop sized jade was

Fig. 4-36 Profound courtyard of Zhang Shiming's Former Residence—Yide Mansion

Fig. 4-37 Hand-cut blue-crystal glass imported from Western countries

Fig. 4-38 Zhang Shiming's Former Residence—Porch in front of Plantain Hall

Fig. 4-40 Zhang Shiming's Former Residence—A western-style dancing hall

Fig. 4-39 Zhang Shiming's Former Residence—Lattice window in Plantain Hall

Fig. 4-41 Zhang Shiming's Former Residence—A western-style building

Fig. 4-42 Zhang Shiming's Former Residence—Castings and balustrades of the balcony at a corner of the courtyard

Fig. 4-43 Exterior of Zhang Shiming's Former Residence in winter

decorated on the small round eye of every plantain leaf (Fig. 4-38, Fig. 4-39). The most amazing part is the two western-style buildings on the west of this architectural congregation. The roofs are in red bricks. The fireplace, glass festoons and Corinthian orders are all in European architectural style of the 18th century. The floor tiles and paintings were imported from France. The bottom level is an extravagant dancing ball, which has a powder room, cloakroom, large fireplace, orchestra pit and crystal chandelier (Fig. 4-40). In the south of the dancing ball, there is a small courtyard in quite different treatment—a few stone columns with capitals carved with the western floral designs supporting a semi-circular balcony railed with balustrades with floral patterns cast in copper. A round mirror is inlaid into the frieze on the top of the roof. The mirror caressed by the leaves and twigs of the southern magnolia has reflection of distant clouds. Facing the small courtyard is a western building with window shutters. The red bricks on the upper part of the gate tower are decorated with the traditional classical patterns of French Toulouse style (Fig. 4-41 to Fig. 4-43). Exclaimed by stunned visitors, this building is no small lakeside town of Taihu Lake but a garden dedicated to Romeo and Juliet! The relish emitted by the manor unfurls the original owner's sophisticated cultural experience and personal preference along with his status, personage, wealth and admiral of the western life style and technological culture.

A quote from an international traveler says, "this is the best, largest, most beautiful and impressive private residence in Jiangnan of China. This manor is a tome." Indeed, Yide Mansion unveils the composed and courageous reaction of a family clan in Jiangnan in response to the inflow of the historical tide of westernization by incorporating unusually touches in the architectural integration of Chinese and Western to their estate with the wisdom of an intellectual businessman. In 2001, Yide Mansion was named "Nanxun Zhang's Old Residence Cluster" and designated as a historical monument and cultural relic under state protection.

4.3 "Western-Style Façade" and "China's Baroque Architecture"

A society has an inclination to relate selectively ideological inspirations from an appealing culture. European Baroque, as an architectural culture appreciated by modern Chinese, during its acceptance and acculturation in China, was blended with the traditional Chinese architecture to become "China's Baroque", which was once the mainstream of Chinese modern "western-style" buildings. Examples in this genre include the Office Building of the Department of Army of the Qing Dynasty (1908—1910), the main gate of Beijing Experimental Farm, the Grand View Tower (Changuanlou) (1906), Shanghai Chengzhong Middle School (1916), Harbin Qiulin Firm (1904), Daowai District Red Cross Hospital (1916), Hankou Lighting Company (1905) and Beijing Ruifuxiang Silk Store (1900), covering building types of office, business, hospital, school, landscape and garden. Among them, the business type was the most welcomed by merchants favoring the "western style façade" to attract patronage that populated cities and towns in the region with shops faced with the bold and exaggerated western Baroque relish. The fad spread into villages and small towns, and the indigenous had integrated seamlessly the Chinese wood imitating structure gate houses with the Baroque style, including even the most traditional ancestral shrine architecture. The variant of the Baroque architecture satisfied the mentality of novelty aspiration and its rich decors resembling the traditional Chinese architectural styles were embraced well during China's modernization. Some of the Baroque vocabulary were translated into a new architecture form through Chinese craftsmanship, which is attested in the creativity and adaptability of "China's Baroque" style.

Fig. 4-44 Exterior of Beijing Ruifuxiang Silk Store in old days

Case 1 Beijing Ruifuxiang Silk Store

Under the trespassing threat of the western power right before and after the 20th century in the late Qing Dynasty, Chinese society and cities reacted with Chinese cultural traits of forgiveness, starting embracing the infiltration of foreign cultures by assertively acceptance or by helplessly passive adaption. Amid the reactions in architecture, many commercial buildings, from style to interior and exterior decorations, were used as the media to vent the dilemma arising from the cultural collision. Some, taking the stance of eclecticism, chose to patch with architectural details, resulting in the typical hybrid style that appeared in China's semi-colony and semi-feudal society at that time. One of them is the reputed old China silk brand—Ruifuxiang Silk Store (Fig. 4-44, Fig. 4-45).

The old address of Ruifuxiang Silk Store was at the core section of the traditional Beijing commercial district Dashila. It is designated as a centennial Major Historical and Cultural Site Protected at the National Level. The store was open in the 19th year of Guangxu reign of the Qing Dynasty (1893). Its current style and appearance were of the rebuilt after a blazing in 1900. The overall architecture is in a traditional courtyard setting, while, partially, it is in the western style with symbolic add-ons. It serves as an exemplar of private Chinese buildings with foreign architectural influence.

Fig. 4-45 Exterior of Beijing Ruifuxiang Silk Store at present days

Fig. 4-46 Ruifuxiang Gate tower in an integrated style of Chinese and Western

The building is based on a brick-and-wood structure in a shop layout with amenities. The shop front is in rectangular shape aligned northward from the south. The building has its roof in a traditional roof style (humpbacked roof with gable), structure in wood, and walls in bricks. The façade facing south is in a typical European Baroque architectural style finished with an recessed arch. A pair of variant Ionic orders at the center frame out the entrance. The shaft and wall are built with deep green bluestones from the west of Beijing. The gate tower is covered with iron sheets. There are five marble tablets with reliefs of traditional Chinese auspicious motifs, such as"Pine and Crane Bless Longevity", "Peonies Present Wealth and Nobility" and "Lotuses Praise Integrity and Virtue" (Fig. 4-46). This kind of complicated hybrid style of Chinese and Western façade used by the Ruifuxiang Silk Store provides appealing appearance and distinct feature. It is an exemplar of China's new shop architectural genre of the 20th century. Behind the entrance gate, there was originally a two-story high courtyard setting used for carriage stop and hitching horses. Now, it is capped as a vestibule (Fig. 4-47) decorated with two delicate brick carvings of "Blossoming Flowers Deliver Fortune" and "Five Blessings for Birthday" made a century ago on its plastered brick sidewalls. In the front, there are an annatto table and armchairs for customer reception

Fig. 4-47 A courtyard in Ruifuxiang Silk Store

Fig. 4-48 The second gate with its architraves decorated with color floral carvings in Ruifuxiang Silk Store

Fig. 4-49 Brick carvings at two sides of the courtyard in Ruifuxiang Silk Store

Fig. 4-50 Courtyard in the business hall of Ruifuxiang Silk Store

Fig. 4-51 Above the courtyard in the business hall of Ruifuxiang Silk Store

Fig. 4-52 Interior decorations with antique flavor in Ruifuxiang Silk Store

Fig. 4-53 Interior of the business hall in Ruifuxiang Silk Store

purpose (Fig. 4-48). At the end of the courtyard, there is the second gate of a pavilion-type with its architraves, door leaves and rafter ends and balustrades decorated with color floral carvings (Fig. 4-49). Today's Ruifuxiang Silk Store has a business section over 1,000 square meters, with its decorations and partitions retained the same as a century ago. Thirty copper foiled columns with carvings and paintings connect to the original building frame. Palace lanterns are used to light the whole place. The ceiling is color painted. The glimmering floor tiles on the second floor were imported from Spain when the building was built. The courtyard at the center of the shop has a sunroof to let natural lights enter the shop for customers to scrutinize the color of the silk (Fig. 4-50 to Fig. 4-53).

Even today, the commercial strip of Beijing Qianmen Dashila is still bustling and hustling and swarmed by customers as a century ago. The silk store

Fig. 4-54 Harbin Daowai District China's Baroque Architecture (from Liu Songfu, Harbin Institute of Technology)

Fig. 4-55 Architecture of conventional courtyard type (from Liu Songfu, Harbin Institute of Technology)

Fig. 4-56 Interior of a courtyard (from Liu Songfu, Harbin Institute of Technology)

has also become a landmark of the strip. In 2006, it was selected into "China's Centennial Architecture Classics".

Case 2 Harbin Daowai District China's Baroque Architecture

In modern Harbin, Daowai District has been an acculturative region. The local craftsmen selectively extracted foreign architectural elements, especially, from the traditional Russian to create a complicated and exceptional architectural genre, "China's Baroque" (Fig. 4-54).

Residential and commercial buildings of mid-sized and small-business shops are the most common style of China's Baroque in this district. Most of the buildings are two or three stories, "front shop and rear courtyard" type. The bottom level is the shop. The second level is the brick-and-wood structured residence. The plan layout is the conventional courtyard type, including Sanheyuan, Siheyuan (quadrangle), multi-courtyard and combined courtyards. The rooms surrounding the courtyard vary on locations, in styles, spaces and sizes. The inner court has cloisters at the second story with access stairways to connect to different areas. This is a common model of space management in China's Baroque and also the spatial characteristics of the traditional architecture of the Central Plains (or *Zhongyuan*, the area on the lower reaches of the Yellow River which formed the cradle of Chinese civilization) in similar architectural integrations of Chinese and Western, fitting into merchants' needs of the front shop and rear courtyard setting (Fig. 4-55, Fig. 4-56).

"China's Baroque" architecture in Harbin Daowai imitates the western-style façade in various styles and

Fig. 4-57 Exterior I of China's Baroque Architecture (from Liu Songfu, Harbin Institute of Technology)

Fig. 4-58 Exterior II of China's Baroque Architecture (from Liu Songfu, Harbin Institute of Technology)

Fig. 4-59 Diverse decoration I of the architectural integration of Chinese and Western (from Liu Songfu, Harbin Institute of Technology)

Fig. 4-60 Diverse decoration II of the architectural integration of Chinese and Western (from Liu Songfu, Harbin Institute of Technology)

Fig. 4-61 Diverse decoration III of the architectural integration of Chinese and Western (from Liu Songfu, Harbin Institute of Technology)

versatile tastes (Fig. 4-57, Fig. 4-58), such as western columns, window casing, friezes with exquisite designs of Chinese craftsmanship in Chinese folkish patterns of pomegranates, grapes, pine, cranes and deer, yearning for prosperous life. These emphases are put on decorations, on lintels, on windows and doors. Windows have shapes in arch, rectangular and oval with rich motifs in volutes, portraits, flowers, twigs, leaves or combinations of the preceding (Fig. 4-59 to Fig. 4-61) decorated on their moldings. Window glass has various aesthetic patterns of infinity, longevity, *ruyi*, and auspicious bat, matching the owner's social status, which are beautiful and delicate, finishing the building with solemnity and elegance. Parapets on the roof top have styles in decorated walls, balustrades, or paneled trellis carved with inscriptions (Fig. 4-62, Fig. 4-63). "China's Baroque" architecture is rich in styles, distinct in layers and exquisite in moldings and details.

"China's Baroque" cluster consisting of several hundreds buildings in Harbin Daowai radiates tenacious vitality and creativity of a marginal culture sprouted from collisions of foreign cultures and indigenous culture. Most of them were constructed by craftsmen of grassroots who believed folkloristic feats. They were the originators and producers of the traditional Chinese culture. Their ingenuities and ideas came from experience of life, which was intuitive and perceptive.

Fig. 4-62 Parapets on the roof of China's Baroque Architecture (from Liu Songfu, Harbin Institute of Technology)

Fig. 4-63 Upper section of China's Baroque Architecture (from Liu Songfu, Harbin Institute of Technology)

Chapter 5 Significance

Historical Significance of Architectural Integration of Chinese and Western

5.1 Acculturation Phenomenon

In the specific modern and contemporary historical period, the architectural integration of Chinese and Western is foreordained to be a cross-culture phenomenon as the modernization movement sprawling over the entire globe landed in China whose literati and peasants oriented society was awakened by foreign influences to engage in remolding itself an industrial and commercial society. Meanwhile, the western missionaries and churches amid their architectural activities were forced to incorporate China's cultural background, sentiment, pressure and spiritual pursues that had shaped those activities to a new landscape into their efforts to not only deal with the cultural conflicts but also carry forward a continuous dialog lasting to present time. Therefore, from an analogous phylogenetics sense, the architectural integration of Chinese and Western is an outcome of cultural conflict and continuous dialog. It can be predicated, as long as the breeding medium of the cultural conflict and continuous dialog actively progressing, changing, and developing, the architectural integration of Chinese and Western will react accordingly until Chinese rebuild up their self-confidence that as the very subject, the "architectural integration of Chinese and Western" will fade to its elimination even if the phenomenon survives. To other countries in East Asia with similar experience, the situation is generally the same. For example, the Imperial Crown style architecture had long been buried in the dust of history, *Datsu-A Ron* (or Escape from Asia and Embrace Westernization) and *Wakon Yōsai* (or Yamato Soul in Westernization) had been wiped out of any dialog, and the cultural conflict had returned to dust. This does not mean the architectural dialog and exchange being stopped between Japanese and the West but the attention has been upgraded from preserving traditional heritage and nationalistic expression to exchange selectively at more equal stance. This has to be attributed to Japan's rise to the second international economic power after Meiji Restoration– a respect earned by self- esteem of a nation and its people backed by wealth and power.

As "carrying forward the orthodoxies of Chinese and the Occident with the cohesion of the essences of both" the core ideology of the architectural integration of Chinese and Western, it is easier to realize than execution. The aforementioned three aspects of the architectural integration of Chinese and Western involve too many disciplines, after a century's continuous digging, no conclusive answer has been found—no consensus has been reached on any theory, approach or implementation by all. However, still facing the challenge, Chinese architectural industry has been tottering along in searching for the mainstream of modernization and has derived several practical issues: How to perceive the nature of tradition? How to comprehend and express critical regionalism in China? How to identify "Chinese Elements"? Although still closely connected to the coarse perception of "China / the West" and "Tradition / Modern", the attempt to remove the fetter of the ideologically driven civilization collision, observe and reinterpret from a spherical perspective the architectural integration of Chinese and Western phenomenon, regardless from the glorious idealism of the rise of Chinese nation or from the realistic industrial bout, the architectural integration of Chinese and Western may be one of the stimuli of China's architectural modernization. This has been realized and implemented in other industries as well, for example, by Li An in movie industry and by Mo Yan in literature, rather than that in architecture.

5.2 Continuity

Wang Shu becoming a Pritzker Architecture Laureate signals a message from the international architectural community of the recognition of the "architectural integration of Chinese and Western" style, or at least, a hint to some insightful inference:

5.2.1 Reinterpreting China Problems

"This is an epoch-making leap. The final

decision to laureate a Chinese architect by the jury marks the Chinese architecture advancement with international recognition. Moreover, the success of China's urbanization as that of the rest of the world in next several decades is crucial globally. It must incorporate the regional specific needs and ethos in its urbanization planning and design in concert with China's long-standing tradition, needless to say the developmental sustainability, in facing this unprecedented opportunity." [1]

Pritzker, when announced the award in person, made the above statement that should have raised close attention of all participants engaging in Chinese architectural activities. The rationale is conspicuous in considering China's rapid economic growth in our modern times. To the world, China's change is no longer nonessential, like any public figure under the spotlight that any gesture will draw heed and criticism. Wang's core ideology sparks off right from the crux of "China's Problems" of our time and that now is a world issue (Fig. 5-1 to Fig. 5-4).

5.2.2 Transforming Historical Elements

"His buildings reveal unique originality recalling

Fig. 5-1 Exterior of a mountain house, Xiangshan School, China Academy of Art (from Zheng Yilin, Zhejiang University)

1 www.pritzkerprize.cn, the Pritzker Architecture Prize (China)

Fig. 5-2 Bird eye's view of No.9 Building, Xiangshan School, China Academy of Art

the past without using the historical elements...Wang's buildings are reputed for its strong sense of cultural heritage and tradition reinterpretation...in his works, history is given a new life as exposing a connection between the past and the present moment. Defining a proper relationship between the past and present is a critical issue since China's urbanization has raised discussions on architectural orientation to maintain the coherence with the tradition or the future. Like all great masterpieces of architecture, Wang's work surmounts arguments and erects as a world monument connecting time and space and buttressed by the

Fig. 5-3 Adytum of a mountain house, Xiangshan School, China Academy of Art (from Zheng Yilin, Zhejiang University)

Fig. 5-4 Interior of a mountain house, Xiangshan School, China Academy of Art (from Zheng Yilin, Zhejiang University)

Fig. 5-5 Ginling Women's College of Arts and Sciences in Nanjing

Fig. 5-6 The Chiang Kai-shek Memorial Hall in Taipei (from Guan Hua, Nanjing University)

Fig. 5-7 Laundries hung in the yard of a Lilong-styled dwelling (built in the Republic of China era) in Meiyuan New Village, Nanjing

historical background."[1]

Here is an implication of valuation: The idea of copying directly historical elements is annulled (Fig. 5-5, Fig. 5-6). Architect as a profession must be able to innovatively solve a specific fundamental problem at a specific time before being appreciated. Using directly historical elements is a lazy solution with semiologic design approach, which can be easily implemented by a civil engineer or a scenic artist, carrying no architectural and real value. In architectonics, it disconnects the traditional style from the modern technology and creates a paradox, negating the architectural performance in society, economic value and environmental ecstasy. Therefore, transcending the tugging between "tradition/future" and ideological shackles of historical elements deserve pursuing. "Atmosphere" is the focus.

5.2.3 Inspiration of Architecture

"I want to build a small city with its own life that can regain conscious". "To me, the fundamental of architecture is spontaneous as a reaction to daily life. When I'm working I am thinking about a 'house' not a 'building'. I relate it to matters in life, things easily being neglected. I named my studio 'Amateur Architecture Studio' to remind my works' spontaneity and experimentation, beyond the "architectonics" [2].

What's noticeable is Wang's intentionally dodging the word "design" but inclining to use words of "build" and "house", reproving and defying problems in contemporary architecture and architecture teaching. Coping, patching, drawing production, GDP appreciation and case study are for beginners, otherwise grotesque and dangerous for professional architects. Taking an analogy in art creation, vibrant arts are usually created from life related subjects – "For so many years, I have been considering architecture as a distinguished subject with etherealness, untouched

1 www.pritzkerprize.cn, the Pritzker Architecture Prize (China)

2 www.jianshe99.com, a website for construction engineering education

virginity and purity. I see architecture in ordinary houses. Under my cognition, a house is more than just a fine plan (design) that is ended as a beautiful picture, rather, an experience of things full of our lives enriched by unpredictability"[1]—a lamentation from Fernando Távora, teacher of Alvaro Siza Vieira (a Portuguese architect, 1992 Pritzker Prize Laureate) after years rich experience from his professional architect career and rumination of self discovery. The professed originality does not have to be loud. It may be just some trivial details scratched and skimped by architects: Has it ever been considered by architects the custom to bask clothes and blankets during spring and winter seasons in the Yangtze River Delta region (Fig. 5-7, Fig. 5-8). Who can aver such attentiveness will not produce dwellings with brand new architectural space? That Chinese architects not short in skills and copycatting but shy in honesty and creativity to tackle the reality being too obtuse to be perceptive has raised concern and perplexity. If elements of "earning a living" and "chasing profit" are translated into "climbing the ladder", the future of architect as a profession would be in doubt, so would architecture as an art be nullified, if the discipline to seek for truth, beauty and virtue is interpreted the same.

Fig. 5-8 Laundries and quilts hung on the balcony of a student dormitory building (built in the late 1990s) in a Nanjing-based university

If broadening the perspective into a more macroscopic view of architect history in Chinese architectural modernization under the global economic context, the room for option and autonomy of undertaking cultural collision, incorporation and exchange has become narrower. This "phenomenal change" is more complicated, deeper and wider than that in previous two transformations. The fast spreading of high-tech brings the scope, speed, intensity and category of the cultural exchange among counties to an unprecedented scale. Pressed by developed countries' economic power and technology advantage in marketing, promoting and propagating of their mainstream cultures and values, the developing countries trapped as latecomers of "exogenous modernization" are forced to face oppression aphasia or marginalization in defending their history and culture under the globalization process. Wang's effort in redefining architecture provides an opportunity, at least, and a direction for contemplation. The "architectural integration of Chinese and Western" genre, to decouple itself from the metaphysical dilemma, such as corporeal/incorporeal, tradition/modern, China/the West, must reclaim the sovereignty in architectural discipline and restore the professionalism of the architect, or the leverage point of the two – construction.

1 Cai Kaizhen, & Wang Jianguo (Ed.) (2005), Alvaro Siza (pp 12), Beijing: China Architecture & Building Press

Sources of Illustrations *

Figures	Sources
Fig. 1-1	Chinese Architecture, 1934, 2(5)
Fig. 1-3	Lu Haiming, & Fan Yi (2012), *Old Nanjing in Postcards*, Nanjing: Nanjing Press
Fig. 1-5, Fig. 2-54 to Fig. 2-56	www.ypdj.cn
Fig. 1-6	Chinese Architecture, 1934. 2(4)
Fig. 1-8	Li Baihao (Ed.) (2005), *Modern Architecture in Hubei*, Beijing: China Architecture & Building Press
Fig. 1-14	*East Changchun Garden of the Gardens of Perfect Brightness* (1931), Northeast Museum
Fig. 1-18	*Lin Keming: A Known Chinese Architect* (1990), Editorial Board, Beijing: Science Press
Fig. 1-19, Fig. 1-20, Fig. 2-10, Fig. 2-72	Lu Haiming, & Wang Xueyan (2012), *Old Nanjing In Photos*, Nanjing: Nanjing Press
Fig. 1-22	*Architectural Design Works of Yang Tingbao* (1983), Institute for Architectural Research, Nanjing Institute of Technology, Beijing: China Architecture & Building Press
Fig. 1-23, Fig. 2-127, Fig. 2-139	Wang Jianguo (Ed.) (1997), *Selected Architectural Writings & Works of Yang Tingbao*, Beijing: China Architecture & Building Press
Fig. 1-24	www.ynmg.yn.gov.cn
Fig. 1-25	www.baike.baidu.com
Fig. 1-26, Fig. 1-27, Fig. 2-125	*Ten Years of Architecture* (1959), Academy of Building Research, Ministry of Construction Engineering (Ed.)
Fig. 1-30	www.ivsky.com
Fig. 1-32	www.hkjy.chineseall.cn
Fig. 1-36, Fig. 2-118, Fig. 2-119, Fig. 4-25	www.nipic.com
Fig. 2-2, Fig. 2-6, Fig. 2-51, Fig. 2-79	www.en.m.wikipedia.org
Fig. 2-15, Fig. 2-39	Chinese Architecture, 1933.1(1)
Fig. 2-17	Lu Haiming, & Yang Xinhua (Ed.) (2001), *Nanjing Architecture of the Republic*, Nanjing: Nanjing University Press
Fig. 2-18	www.commons.wikimedia.org
Fig. 2-44 to Fig. 2-48, Fig. 2-50	Li Haiqing, & Liu Jun (2001), *Growing through Exploration—Untold Story on Problems of Old National Central Museum Architecture*, Huazhong Architecture (Issue 6)
Fig. 2-49	Nanjing Museum
Fig. 2-52, Fig. 2-57	*Ten Years of the Department of Public Works* (1937), Department of Public Works of the Shanghai Municipal Government

* Editor's note: the images with sources already provided in the article are not included and the unattributed images are the work of the author.

Fig. 2-53, Fig. 2-58	Chinese Architecture, 1933.1(6)
Fig. 2-61, Fig. 2-62, Fig. 2-69, Fig. 2-70	Chinese Architecture, 1934.2(2)
Fig. 2-63	Architecture Monthly, 1933.1(6)
Fig. 2-73, Fig. 2-74, Fig. 3-8 to Fig. 3-10, Fig. 3-16, Fig. 3-26, Fig. 3-28, Fig. 3-29, Fig. 3-32 to Fig. 3-38	Han Dongqing, & Zhang Tong (2001), *Yang Tingbao Architectural Design Collection*, Beijing: China Architecture & Building Press
Fig. 2-86, Fig. 2-87	Architecture Monthly, 1936.4(2)
Fig. 2-90	www.hi.baidu.com
Fig. 2-101, Fig. 2-102, Fig. 2-107, Fig. 2-109	www.zchqc2.xmu.edu.cn
Fig. 2-106	www.qzrz.org
Fig. 2-124	*Selected Design Works of Teachers* (1987), School of Architecture and Institute for Architectural Research, Nanjing Institute of Technology, Nanjing: Nanjing Institute of Technology Press
Fig. 2-141	www.en.academic.ru
Fig. 2-145	www.ad.ntust.edu.tw
Fig. 2-146, Fig. 2-147	www.forgemind.net
Fig. 2-148, Fig. 2-149	www.bnw.com.tw
Fig. 2-155	www.dbk2.chinabaike.org
Fig. 2-156	www.pccu.edu.tw
Fig. 3-17	Chinese Architecture, 1935
Fig. 3-18	http://tupian .baike.com
Fig. 3-49	Architecture Weekly (Vol. 3, Issue 5), 1935
Fig. 3-53, Fig. 3-54, Fig. 3-56	Zhang Kaiji et al. (1957), *Beijing Planetarium*, Architectural Journal (Issue 1)
Fig. 3-59, Fig. 3-60, Fig. 3-61	http://www.bjp.org.cn
Fig. 3-63	*Beijing Minzu Hotel* (1959), Minzu Hotel Design Team, Design Institute of Beijing Municipal Administration of Urban Planning, Architectural Journal (Supplement Issue 1)
Fig. 3-69, Fig. 3-70, Fig. 3-71	http://tszyk.bucea.edu.cn
Fig. 4-1	*Charm of the Old Shanghai* (5 volumes) (1998), Shanghai Library (Ed.), Shanghai: Shanghai Culture Publishing House
Fig. 4-2, Fig. 4-3, Fig. 4-4, Fig. 4-5, Fig. 4-12	*Old Alley Jianyeli*(2009), Shanghai Zhangming Architectural Design Firm, Shanghai: Shanghai Far East Publishers
Fig. 4-11	http://tupian.hudong.com
Fig. 4-24	http://guidebook.youtx.com
Fig. 4-28	http://www.yikuaiqu.com
Fig. 4-34, Fig. 4-35, Fig. 4-38	http://baike.baidu.com
Fig. 3-43	http://blog.163.com
Fig. 4-44	http://www.rifuxiang.cn

Index

1. Index of Glossary

2.Index of People

Afterword

About ten years ago, the first time after reading *The Great Vehicle of Architecture* of Mr. Han Baode, I thought I had understood something. I reread now and then afterwards, and I learned something new every time. The contact I have made with the architectural culture started then. Now as I recollect, it is normal. How could a fledgling young person who had been buried in books and drawings most of the time and had never built a house, really catch the meaning of architecture?

After almost 20 years honing of the architecture teaching career, I start realizing the accuracy of the insightful and prudent statement made by Mr. Han, "As an instrument and a manifestation, retaining originality and unpretentiousness, Chinese architecture in nature is a construction of existence, related closely to man. It belongs to the humanistic construction out of intuition." And, "The Chinese have never intended to change the product of intuition into diverse ostensible forms, and have never been shackled by architectural tradition to reject the need of circumstantial modification. Through thousands years, therefore, as Chinese cultural evolves, so does architecture. It is a book of Chinese history. It records how intelligentsia yearns for the settlement of mind, how the ruling class parades its power of supremacy and how the rich and the tycoons indulge themselves in decadent pleasure. The simple close to primitive architectural space framework reveals all—this is a wonder of world's architecture."

It cannot be more apropos to interpret "the architectural integration of Chinese and Western" with this statement. It even allows tolerance for elaboration.

This book would not have even been in its formation if not for the trust and recommendation of Professor Chen Wei, who have been giving me support and help since the decision on my dissertation topic fifteen years ago. However, my slow learning capacity cannot repay the full expectation but continue my endeavor. The theoretical foundation for this book is explicated from the result of years' studying and discussion under and with Professor Zhao Chen. The academic wisdoms and dispositions of the two teachers have been my model to pursue though probably unreachable. In my daily study, discourses with Li Hua, Shi Yonggao and Leng Tian have benefited me and the long time support from my predecessors–Lai Delin, Li Baihao, Peng Nu, Peng Changxin, Tan Gangyi and Feng Jiang, teachers and friends have underpinned the research work of the book. I also appreciate Qin Lin, Guan Hua, Shi Qingchao, Liu Fei and Zhao Yunfei, respectively for their touching photos of architecture in Chongqing, Taipei, Xiamen and Guangzhou. My special thanks to Deputy Editor in Chief Zhang Huizhen of the publisher and Editor Dong Suhua and Qi Linlin, for without the confidence of Zhang and the prudence of two editors, to finish this work not large in scale but substantial in significance would have been unimaginable.

Of course, I cannot forget the consideration and support of four females in my family, my daughter, wife, mother and mother-in-law.

The conundrum discussed in this book is still undergoing its evolution. No conclusion is yet in sight, which makes it more attractive. Any persuasions in any forms are welcome with expectations.

Li Haiqing
Jinling Banshan Ju, July 2014

Li Haiqing

Born in Anhui in 1970, Mr. Li Haiqing has a Ph.D in Engineering and works as an associate professor and a graduate student advisor in the School of Architecture of Southeast University. Li has been engaging in the studies of the technology history of modern Chinese architecture, indigenization design strategy for green buildings, and exploration of construction conception of contemporary Chinese architecture.

In 2004, he published the treatise titled "Study on Modern Transformation of Chinese Architecture", through which he conducted deep investigation from a unique perspective on contemporary development of Chinese architecture. In recent years, he has been devoted to the theoretical research of how to build indigenous architecture of China. In addition to releasing more than 40 theses in this field, he has also presided or participated in over 30 important projects including Nanjing Xiao Xian Memorial Hall. Besides, some undergraduates and graduate students under his guidance have won prizes from many architectural design competitions or assignment reviews in China.

Wang Xiaoqian

Born in Yangzhou of Jiangsu, Ms. Wang Xiaoqian has a Ph.D in Engineering and works as an associate professor and a graduate student advisor in the History and Theory Research Institute of the School of Architecture, Southeast University. Wang has been engaging in the studies of the history of world architecture and art, modern Chinese architecture, protection and renovation of architectural heritage, and sustainable living environment design.

So far, she has presided and completed more than 10 scientific research programs at the national, provincial, municipal, school levels as well as over 10 projects involving heritage protection and renovation. In recent years, she has released more than 30 academic papers on China's top-level and core periodicals, edited and written eight books including textbooks, and won a string of prizes such as China National Book Award, China Architectural Book Award, and the Le Corbusier Award by the Architectural History Branch of the Architectural Society of China.

The architectural integration of Chinese and Western, as a value orientation of architecture design, grows in time and serves as the impression for the vibrant puzzle of the dynamic modern transformation of Chinese architecture. It perpetuates its vitality of cultural grafting phenomenon and influence on the contemporary Chinese architecture through architecture activities.

Its historical background is rather unique. Its emergence, blossom and alteration are dramatic. As a cultural heritage, its architectural art value has not been attended and elaborated. This book categorizes accurately the formulation mechanism from the theoretical perspective of architecture. By observing and depicting physical architecture activities, it attempts to sweep panoramically architectural cultural ideologies in modern times and give evaluation for particulars. It sorts out "the architectural integration of Chinese and Western" phenomenon in documents from the Chinese architectural art perspective, followed by using the "contemporarily styled Chinese classical architecture", "Chinese Modern Architecture" and non-professional architecture as media to categorize and introduce architecture with "the architectural integration of Chinese and Western". It envisages a future path for "the architectural integration of Chinese and Western" in its conclusion.

The entire book has clear venation and rich photos and illustrations. It can be used as a reference book and teaching materials for history of Chinese architecture in art colleges. It can also serve as a reading material for aficionados of Chinese architecture and culture.

近世中西合璧建築藝術

The Art of Architectural Integration of Chinese and Western

Written by Li Haiqing, Wang Xiaoqian

Translated by Gary Sun, Song Yan/ China Translation Corporation

General Planner: Zhang Huizhen

Executive Planners: Zhang Huizhen, Qi Linlin, Dong Suhua

Executive Editors: Dong Suhua, Zhang Huizhen, Qi Linlin

English Text Proofreading: He Guangsen

Technical Editors: Zhao Zikuan, Li Jianyun

Contributing Art Editor: Miao Jie

Overall Design: BEIJING BEAUTIFUL GRAPHIC CO., LTD

First Published in 2015 by China Architecture & Building Press

ISBN 978-7-112-17621-2 (26833)

CIP date available on request

www.cabp.com.cn

Printed on acid-free paper produced from chlorine-free pulp. TCF∞
Printed in China
Price: RMB 160